THE EXTRAORDINARY WORLD OF

DIAMONDS

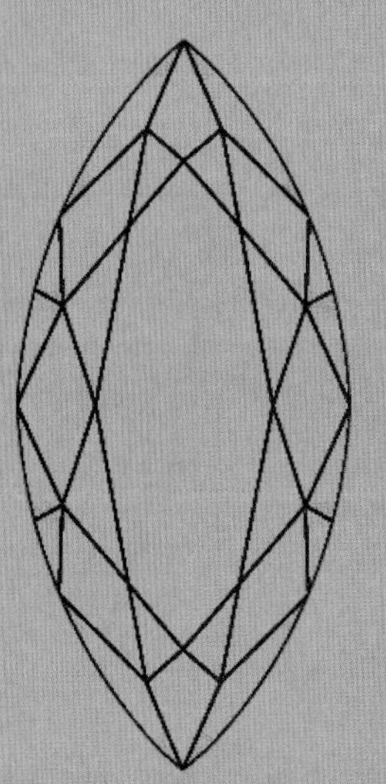

THE EXTRAORDINARY WORLD OF

DIAMONDS

NICK NORMAN

A WORD OF THANKS TO MY PARTNERS

CHRIS JENNINGS

It would be no exaggeration to say that Chris and Jeanne have been with me from the first step of the project, full of counsel, wisdom, and technical and historical information whenever I called for it and, most critical of all, generous funding. I certainly could not have done the book without you. The financial support came when the shape of the book was barely defined and with nothing like a contract to secure it: this was, as a complete act of faith, a huge compliment to me. If you, who have given your life to diamonds, can feel a measure of pride to be associated with this endeavour, it is small compensation for the honour you have done me.

E. OPPENHEIMER & SON

The research for the historical part of a book like this does not proceed very far before one comes across the name of Oppenheimer. For a hundred years it has resonated through the world of diamonds. The name has provided stability in what was until recently a precarious business, and both the industry and wearers of diamonds owe them a debt of thanks. Mine is for recognition of my humble book by the family's sponsorship. Thank you.

THE OPPENHEIMER MEMORIAL TRUST

The motivation to write a book like this comes from a sense of wonder at the world we live in, so perfectly symbolised by the most miraculous mineral we know. That sense is a reminder of the privilege people like me have enjoyed in our boundless education, and spurs one to reciprocate in some small way by opening that world to others. Organisations like OMT that seek to find ways of reaching out in educational projects deserve the thanks not only of the recipients: I thank you for the opportunity to join the family of contributors.

DE BEERS

A company that dominates an industry for a hundred years accumulates not only great gravitas, but a comprehensive record of its progress. The written assurance I received from you of your technical and moral support deferred to the former; from the latter came a stream of photographs without which the book would have been a paltry shadow of the vivid illustration in your hands of a world known to only a lucky few. I salute you: I could not have done it without you.

THE GEOLOGICAL SOCIETY OF SOUTH AFRICA

One of the proudest credits in my curriculum vitae is my fellowship of this venerable institution. My thanks go to you and your management for support of my various endeavours, exemplifying as it does your ethos of continually exploring new ways to facilitate projects that make the geosciences more accessible to the world at large. May this book contribute to that goal.

CONTENTS

FOREWORD

Nick Norman's book starts with a bang and then proceeds to captivate the reader until the very end.

His book is a rare combination of superb technical knowledge, clearly interpreted for both professionals and laymen, and lucid prose, interspersed with classical mythology and quotations from classical literature, which makes for excellent reading. He takes his readers on an exciting journey through all stages of the diamond industry as well as through the history of diamonds, from the first discoveries in India thousands of years ago to those happening in Canada today. The amount of research carried out for the book is illustrated in the depth and breadth of his coverage of this intriguing subject.

I first met Nick forty years ago and our friendship has survived living on different continents and in different hemispheres. We both spent many happy years in Botswana as geologists and it was probably there that we both developed a passion for diamonds as objects of unparalleled beauty presenting the huge challenge of understanding and revealing their origins to those wishing to find them. It has been my privilege to know Nick as a thorough gentleman, an accomplished geologist and talented writer.

I feel sure that this book will become a standard reference for all who are fascinated by this most extraordinary of gemstones.

Chris Jennings
Grand Cayman, Cayman Islands

PREFACE

Gloria in excelsis Deo

'A diamond is forever', De Beers tells us. Ian Fleming corrupted the by-line but not the message in the title of an early James Bond novel, and Marilyn Monroe sang famously that 'diamonds are a girl's best friend.' There are those who say these are glittering facets of a gigantic hoax: to them, as to anyone who cares to listen, I say there is a far better reason for the hold diamonds have on us than the perpetuation of a myth by De Beers. There is nothing mythical about the uniqueness of the diamond.

If life is one miracle that carbon gives us, the other is the diamond, a stone that should never have made it far up the pounding Atlantic beaches of Africa because it should not have reached the surface at all. No other gem that graces hands and throats was forged hundreds of kilometres below the surface from a single simple element, at pressure and temperature almost beyond conception. Its simplicity and symmetry have given it fire and life unmatched by any other mineral. They imbue it, too, with hardness far in excess of other gems, without which it could not travel thousands of kilometres in raging boulder-laden rivers and up one of the wildest coasts in the world.

Marvel at the greater parts of the miracle: its creation and transport to the surface. Only in exceptionally pressurised nuclei in Earth's hot mantle will carbon not crystallise as graphite, as soft as the diamond is hard. Having formed, the diamond must come up very fast to survive to the surface without reverting to its malleable form or going up in a puff of carbon dioxide. Like an express train leaving the station, the kimberlite that transports the diamond starts slowly, steadily gathering speed until, bullet-like, it reaches the daylight, leaving the diamond no time to convert to graphite or oxidise. Think of it: if diamond was stable anywhere near the surface, alchemists and scientists would not have needed hundreds of years to crack the technology of how to synthesise it.

Were diamonds rarer than any other gem it might be said that their value lay in th
scarcity. That is an artificial value, though, and, contrary to a school of thought that verge
cynical, there is nothing artificial about the value of the millions of natural diamonds, b
small, that are proudly owned around the world. Though not rare in nature, from the
of Western Australia to the Skeleton Coast of Namibia to the steppes of Siberia, the
leanly spread through their host rock than any other mineral mined. Finding th

If the route taken by the diamond from its formation in the depths onto
makes a fascinating story, how much more so the narrative of the unfoldin
the characters, the intrigue, the drama, the romance. These are what thi
gleaned from the hard ore of history and anecdote to entrance you, like
that tumble onto the sort-house conveyor belt.

ACKNOWLEDGEMENTS

This little book took over three and a half years of hard work to do – not quite blood, sweat and tears, but countless hours at a computer – writing, researching and sourcing visual material. It was not a case of 'I did it': far from it. It was only by the grace of God, and a massive joint venture, between Him, me and a lot of very important people. Without some of them it simply would not have happened.

Without the incredibly generous financial support of Chris Jennings the book would have been dead in the water. He backed it with an abundance of technical input, wisdom and liberal moral support.

When I was unable to raise the sponsorship needed to travel to the countries covered, I thought my project had become Mission Impossible. Only the unflagging enthusiasm and encouragement of Jenny Hobbs, with a lifetime in books behind her, persuaded me otherwise. She spent hour after hour reading drafts of chapters, correcting, counselling, and always gently but firmly pushing. Another Chris was critical, this one Chris Smith. Although I still have not met him, he has been a constant source of technical information and material integral to the book and inaccessible through normal channels. The chapters on Russia, India and particularly Australia would have been pitiful offerings without massive input from him, aided and abetted by his Russian geologist wife, Galina Bulanova. Together they translated many pages of Russian documents and put me in touch with key people in the diamond industry there. Chris is central to the Argyle discovery story: getting that from the horse's mouth is one of the aspects that make this book unique. The same applies to Chris Jennings in the Canadian narrative.

Writing a book on diamonds in a place far from the diamond fields would have been illogical but for the fact that one of the revered pioneers of the science of kimberlites and diamonds, John Gurney, has his consulting business in Cape Town, an hour's drive away from where I live and write. His enthusiastic support and advice have been indispensable, and no less so the ever-willing availability of associates Jeanine, Kendra and particularly Megan to answer questions and locate technical material. Thanks, too, to Tom Nowicki of Mineral Service's Vancouver office for technical input.

Before listing all the colleagues who have given willingly of their time, most of them lots of it, I should single out a few friends without whose belief in the project I might have wavered. Ex-De Beers managing director Gary Ralfe in the beginning showed no doubt that the project was worth doing and made useful contacts for me in De Beers. James Allan, as a diamond industry analyst of high repute, was generous with time, support and advice and made important introductions for me. John Rudd, with an almost unparalleled depth and breadth of experience in the industry, willingly put himself and his library at my disposal, and, thanks largely to his ex-wife Anna de la Garda, I was lucky enough to inherit some extremely useful books when he died, about half-way through the project. Baxter Brown, another widely respected veteran of the diamond business, was always at the end of a telephone line to answer questions, as well as lending me technical papers that I would otherwise not have known existed. Fourth-generation diamond digger *xtraordinaire* Dick Barker gave freely of his time, expertise and hospitality and added richly to y diamond experience; and his son, Maurice, also well versed in diamond technology, became riend whom I called on often for help.

Keith Whitelock drove me to the Letšeng mine and hosted me there for as long as I needed, and I warmly thank him, as well as Ray Ferraris and Debbie Bowen for setting up the visit and showing me some beautiful stones. Ann Snaddon deserves special mention for making available (through her daughter Jiggs) the diary of her mother, who as a surprised five-year-old held the Cullinan diamond in her 'very small paw'.

In the Oppenheimer 'family', Diana Madden at the Brenthurst Library found a copious variety of archival photographs for me to draw on, and Clare Digby of the Oppenheimer Memorial Trust listened sympathetically to my pleas for financial assistance, as did Kalim Rajab, Nicky Oppenheimer and E. Oppenheimer & Son's Duncan Macfadyen. Without their provision the book would not have been as sumptuously illustrated as it is.

Many of the photographs in the book are from the De Beers stock, provided free of charge. There special thanks are due to photo-archivist Marlaine Botha, for whom no request was too much trouble. The access was authorised by communications manager Tom Tweedy. To those individuals, as well as the corporate entity of De Beers: thank you. Burger Greeff and Kevin Richardson of De Beers Marine gave me an insight into how diamonds are recovered off the seabed and I thank them. Two ex-De Beers exploration gurus, Barry Hawthorne and Bill McKechnie, are gratefully acknowledged for time freely given. Ian Corbett, ultimate authority on wind-concentration of diamonds from years in the Namib's stormy blast, could not have been more helpful and I am deeply grateful to him. Various people in De Beers in London were instrumental in getting me a 'letter of comfort'. Thank you.

Thanks are due, too, to Gill Lanham, who worked tirelessly to source visual material from image libraries, and to Elmi Pretorius, who made the maps and drawings to a very high standard. Photographer Trish Heywood was available for various bizarre missions and she is thankfully saluted. Russell Martin, Jacana's Cape Town director and editor, had to work hard to contain my tendency towards verbosity and at times a plethora of irrelevant detail: you owe him as much thanks as I do. Towards the end he was a victim of my disorganisation as loose ends were being tied together, and I am grateful to him and to designer Abdul Amien for their persistence in putting together something I hope we can all be proud of.

Others I have prevailed on over the years for input of a wide variety are Chris Morrissey, Craig Smith, Victor Masaitis, Victor Ustinov, Vladimir Sklyarov, Vladimir Shchukin, Nick Sobolev, Guy Pas, Paul Sobie, Harrison Cookenboo, Leila Benitez, Guilherme Gonzaga, Darcy Svisero, Ken Johnson, Cathy Rodrigues, Luiz Claudio Marigo, Atmavireshwar Sthapak, John Collier, Warren Atkinson, Ewen Tyler, Peter Kennewell, Mick Paltridge, Edward Gajdel, Marion Bamford, Sherryn Tedder, Cathy Roberts, Jim Davidson, Greg Stephenson, Philip Mostert, Mike de Wit, Phillida Brooke Simons, John Lincoln, Fleur Way Jones, Clive Hassall, Hazel Fraser, Jay Barton, Thomas Branch, Cezar Ferreira, Blake Flemington, Joppie Nieman, Leané Prinsloo, Susan Winckler, Chris Hellinger, Anton le Roex and Hugh Leggatt.

And last but by no means least, thanks are due to Sue. All but a few writers know that writing is a passion not a job, where your time could be invested far more profitably doing real work, earning real money. To be able to do it, you need some other form of income: Sue has worked night and day to make it possible for me to write. And Sarah and Christopher have had to endure a father whose stress levels made him at best grumpy, mostly irrationally irritable. Thanks, you two, for not leaving home.

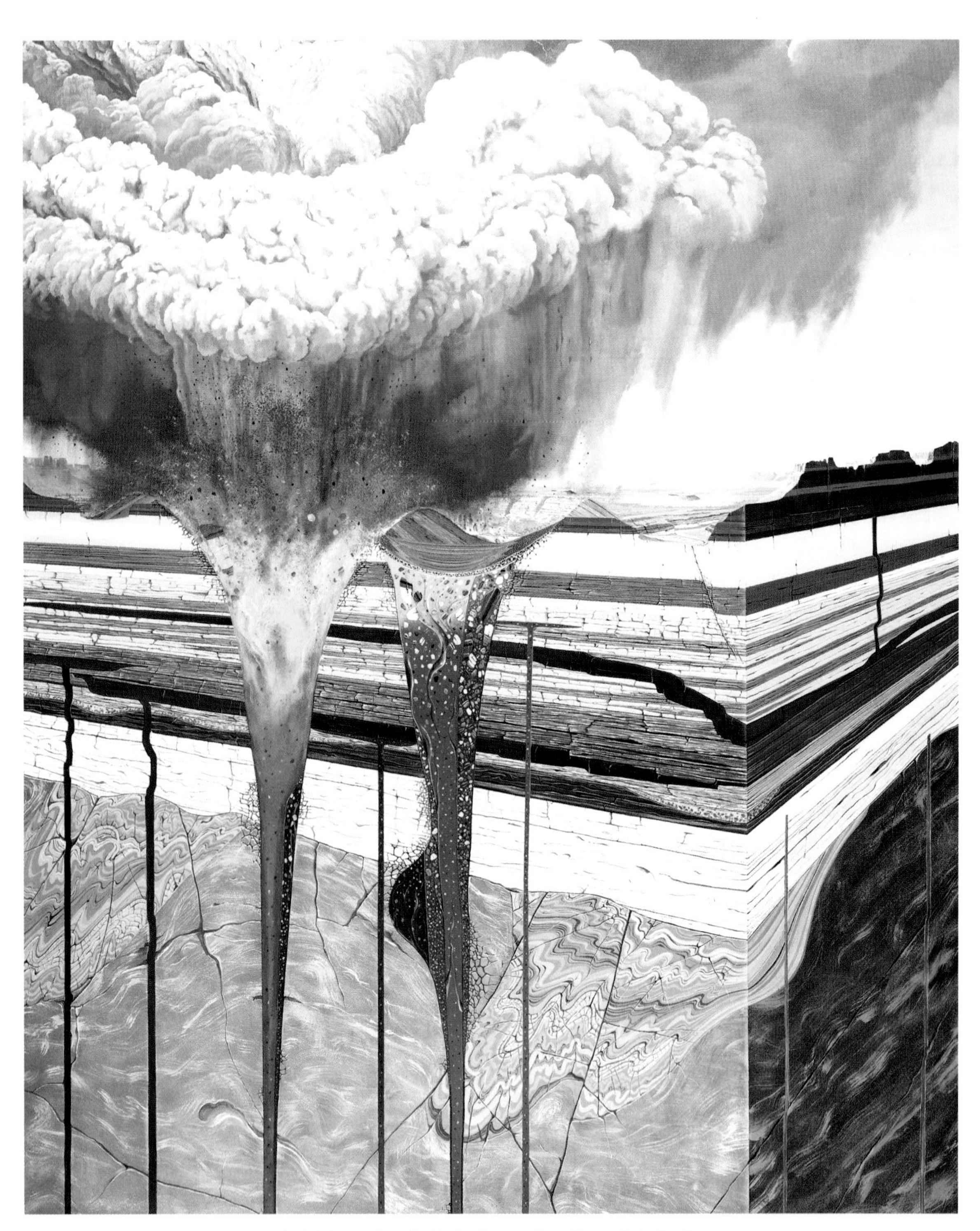

Artist's impression of a kimberlite eruption: diamonds in the sky.

CHAPTER 1
INTRODUCTION

From top to bottom the Earth's crust shuddered in a brief spasm. On the surface Nqwebasaurus, a lizard-sized dinosaur, stopped in his tracks as the grass around him stirred on a breathless African afternoon. He moved onto a stony mound and scanned his world in all directions. A loud rumble from deep below turned into a deafening roar as a black column of rock, from blocks bigger than he was to fine dust, broke through the surface and was blasted high into the sky. The explosion was closer to the horizon than to where he stood, yet the sun was blotted out. Gradually the shadow lightened as the stuff fell, most of it back into the opening the blast had made, some of it onto the ground around the vent, a raised rim. The cloud, now dissipating, looked black from where he was, except that once or twice, as he watched, he saw its darkness pierced by a tiny bright star catching the sun's rays for an instant as it fell. The fine dust settled slowly, and after a while Nqwebasaurus continued on his way. He had witnessed a kimberlite eruption: diamonds in the sky.

Gaius Plinius Secundus – better known as Pliny the Elder – was not so lucky. In AD 79, Pliny helped evacuate residents of the towns around erupting Vesuvius. Prevented by volcanic debris from getting as close as he would have liked to the threatened towns, Pliny landed at Stabiae, 16 kilometres from the eruption vent. He died the following day, in a cruel twist of fate, himself a victim of the most violent eruption of the series.

Of diamond, or *adamas* in Greek (cf the English word 'adamantine'), Pliny, a naturalist, had this to say, 'The substance that possesses the greatest value, not only among precious stones, but of all human possessions, is adamas.' When he penned the words for which he is best known, 'Ex Africa semper aliquid novi', he was making a connection that would manifest itself eighteen hundred years later.

In 1871 diamonds were discovered near Colesberg Kopje, in the centre of South Africa. They were not the first to be discovered in that country, but what makes them historically pivotal is that they were the first to be discovered where no river had brought them, in the 'mother lode'. By the time a sizeable town had sprung up next to the diggings, two years later, the place was named Kimberley, after the British Secretary of State for the Colonies, the Earl of Kimberley. It was not long before immortality came to the Colonial Secretary in a way he could never have dreamt of: the mother lode – the so-called blue ground – found its way into the geological lexicon as 'kimberlite'.

If Pliny was tens of millions of years too late to see a kimberlite pipe erupt, he was two thousand too early to see the rock that brought his *adamas* to the surface. And if diamonds were discovered in South Africa in a time so recent as to have been captured on film, on a continent mostly unknown to the civilisations of Greece and Rome, where did these stones which he prized so highly come from?

They came from a country with an ancient recorded history, India, where their existence had been known for at least six hundred years. From the beginning the east-to-west trading route has carried not only the spices that Western palates relish but a sprinkling of fabulous gems besides. If the royal regalia of Indian potentates was adorned by the best that could be dug from the gravels of Golconda, Persian shahs and European princes were not to be upstaged.

Sayajirao Gaekwad III, Maharaja of Baroda, in northwestern India, epitomised the taste that Indian potentates acquired for diamonds.

That two and a half thousand years separate the first diamond rushes of India and South Africa is of interest to a handful of historians, but not to geologists. Their interest stretches back billions of years to when the crystals first began to form: to a time when there was no ocean separating the two coastlines that today face each other across the equator; to a time long before the fragmentation of the ancient supercontinent, Gondwana; before the collision between India, coming from the south, and Asia, far to the north, gave us a welt of buckled-up crust that towers over the rest of the planet, the Himalayas; to a time when the stable, deep-rooted nuclei of Africa and India shared the same neighbourhood in Earth's lithosphere. That is the time when – for diamonds – it all began.

In Africa there are several such nuclei – geologists call them cratons – still preserved. They were all penetrated at one time or another by kimberlite pipes, the volcanic feeders that shot the diamonds from their source deep in the mantle to the surface. The West African cratons had a neighbour, too, which moved away across an ocean. The São Francisco craton in Brazil has distinguished itself by giving us diamonds, like those of India and Africa, illustrious enough to have earned names for themselves: among them the President Vargas, standing proud next to the Kohinoor and the Cullinan. Until the closing days of the second millennium, these three countries had produced the great majority of gem diamonds. In the history of diamonds, Brazil lies sandwiched between ancient India and modern South Africa.

By the late eighteenth century Europe's rich and famous had acquired a taste for expensive baubles. The growth of interest was ill timed, though, for production from the alluvial fields of Golconda was dwindling. Gloom threatened. But good news was on the way via Portugal:

in the Brazilian state of Minas Gerais, already an important gold producer and a veritable Aladdin's chest of semi-precious stones, river diamonds had been found. And as European eyes turned westwards, those of the Brazilian prospectors, or *garimpeiros*, acquired a new focus. Now it was not only the tail of gold in a pan that drove them to extremes of hardship, it was also the shining pale crystal at the apex of a sieved heap of concentrate.

But the boom days in the eastern savannah of Brazil were short-lived. After a hundred years of free access to the European markets, the New World source was to be eclipsed. For in 1866 a frontier farmer on the Cape Colony's Orange River near Hopetown was offered a shiny stone that had drawn his attention as a child's plaything. Hardly believing his luck, he accepted it and gave it to the local trader when he next passed that way, with the request to have it identified. Confirmation from a doctor turned amateur geologist in Grahamstown that it was indeed a diamond stirred little excitement among pioneer prospectors, even less in the European markets; and for two years it was mainly farmers and their shepherds who picked up occasional small stones near the river. A steady stream of Brazilian stones was still reaching the polishing wheels and jewellery settings in Amsterdam and Antwerp, where diamonds from the Dark Continent to the south barely raised an eyebrow. It would take the 'dry diggings' around Colesberg Kopje in 1871 to change all that.

In March 1869 the discovery of an 83-carat stone on the Orange River was reported in the press, triggering a rush that within weeks would see rough tented camps dotted all along the Orange River and beyond. Richer pickings soon came to light along the Vaal River, a few days' trek upstream from where it joins the Orange, and the camp of makeshift structures that came to be known as Barkly West was in a short time the centre of a helter-skelter rush. From there the fever spread; across the African veld eyes were turned downwards, for a sign, a glint, on the dusty paths.

The Eureka Diamond, the first to be found in the Cape, in 1866.

No one could guess the impact of the first 'dry diggings', an easy day's ride east of Barkly West. Far from any stream or river, diamonds had been found in yellow clay. With none of the boulders and pebbles that blunted picks on the river diggings, the going was quick and easy and, as they dug, the miners found a plentiful supply of gems. Though the quality of the stones

De Beers was already a name in diamond mining in 1902.

may not have been as good as those from the river diggings, their profusion was unheard of. The biggest rush since Barkly West was soon under way, and buyers found themselves working late into the night, assembling parcels for dispatch to markets a world away.

Now the European merchants, or diamantaires, took notice. Not only was a supply reaching them as never before, but it included some of the finest stones they had ever seen. The diamond traffic was breaking out of the fetters that had bound it for two thousand years, when all stones came from river channels, far removed from the mother lode. With rich primary deposits located at last, the trade was about to enter a new era.

The key to this largesse lay not only in the discovery of the rock with a new name, kimberlite. Another hitherto unknown name was equally crucial: Cecil John Rhodes, who at the end of a bitter feud with a fellow entrepreneur-miner, Barney Barnato, had established himself as the king of Kimberley by 1888. And if the name kimberlite is well known to geologists, and Cecil Rhodes to historians and the scholarly elite, that of the company Rhodes formed to mine the rock has become a brand name second to none: De Beers or, to give it its full name, De Beers Consolidated Mines Limited.

Since the late nineteenth century, the story of diamonds has been inextricably linked with that of De Beers. Together with the London Diamond Syndicate, formed to market the burgeoning production from the Kimberley mines, De Beers controlled the supply chain of diamonds by the turn of the century. The cartel was born of necessity. For the first time in history the supply of diamonds had become so prolific that it far outstripped demand. Retailers found new merchandise languishing on the shelves: if they were to drop the prices any further, they would be selling at a loss. For De Beers and the Syndicate the solution was simple: they would throttle the flow into the market. The rate of mining was slowed and purchases from diggers all but stopped. Imperceptibly, as the European showcases became less cluttered, the scales began to tip. Slowly, ever so watchfully, stockpiled stones were released into the market. But diggers and mine managers knew that the honeymoon was over.

The next blow De Beers was dealt would prove almost fatal. When Cecil Rhodes died in March 1902, just short of his fiftieth birthday, the diamond-mining industry was about to be turned on its head by the purchase of a farm just east of Pretoria by a man named Tom Cullinan. The pipe that Cullinan had been sure existed on the farm, first known as the Premier and much later renamed after its discoverer, was as big as all the Kimberley pipes together and soon was producing a regular supply of large stones of the finest quality. With De Beers no longer able to exert the stranglehold on supply necessary to stabilise prices, the industry entered a free market environment. Stones flooded the markets in Antwerp, Amsterdam and London.

And the situation would get worse before it got better. In 1908 the discovery of a diamond by a labourer clearing sand off the railway track near Lüderitz in German South West Africa (now Namibia) precipitated a rush that would open a vast new diamond field, in a foreign land a world away from Kimberley and Pretoria. Soon the supply had become a torrent.

It took the onset of hostilities in Europe in 1914, a few years and a mild-mannered Jewish entrepreneur to restore stability to an industry in turmoil. But we need to turn the clock back, to 1902. A pivotal year for the diamond industry, it not only saw Rhodes's death, and the discovery of the Premier pipe; it also marked the arrival of Ernest Oppenheimer in Kimberley. Twenty-two-year-old Ernest had been sent by his diamond-dealing family in London to look after their interests in the Cape. He quickly filled the niche his uncles had seen for him there and was soon widely respected for his discernment and acumen. But for him, and for now, the real opportunity lay further north. On the Witwatersrand, he showed the entrepreneurship that was to become the hallmark for him and for future generations of Oppenheimers. Using ingenuity and American financing, he set up the Anglo American Corporation, soon to be a powerhouse in the gold mines opening up all along an ever-extending strip of highveld. In an arena of business opportunity without equal anywhere in

The Premier Mine (now the Cullinan Mine) in the early 1900s.

the early twentieth century, the young Ernest Oppenheimer had shown that he was a force to be reckoned with.

Up-and-coming gold baron he might have been, but his passion for diamonds never left him. That fervour, matched with vigilance and quiet persuasiveness, would soon secure for Oppenheimer the Atlantic coast diamonds that the British and their colonial surrogate, South Africa, had made their own with the defeat of the German forces in South West Africa. Oppenheimer incorporated his new acquisition into the Consolidated Diamond Mines Limited. This powerful foothold in the diamond world, and his relentless purchase of De Beers stock, ultimately won Sir Ernest Oppenheimer, now converted to Christianity and knighted for his contribution to the war effort, the chair – and control – of De Beers in 1929. His lifelong goal had been achieved. With the Premier Mine forced by the crippling pre-1914 price war into the arms of De Beers, Oppenheimer's control of diamond mining was all but complete.

Now that he had assumed the running of De Beers, Oppenheimer had one last item of business to attend to in the marketplace. Though not prevalent at the time, chaos at that end of the diamond chain was always a threat. Oppenheimer had seen it once and was determined that it wouldn't happen again. The producers – mainly De Beers, but with West African and

Earliest days on the Witwatersrand, not long before Ernest Oppenheimer arrived.

Brazilian mines by no means inconsequential – were at one end of the chain, the consumers at the other, and between them lay a chasm. It was a void in which stones with an infinite spread of values, and diamantaires with an almost equal range of integrity, teetered between supply and demand with the balance changing daily. Only single-channel marketing of rough – as uncut diamonds are called – would bridge the chasm and prevent the oversupply that had seen prices tumble headlong.

When the players dusted themselves off and their eyes had cleared, they saw that no one, barring the few who had bought as the market bottomed, had gained from the fracas. Oppenheimer reinstated the old syndicate. Though never existing as a commercial entity, the Central Selling Organisation, in Charterhouse Street in the heart of London, would effectively maintain tight control of the supply of rough into the marketplace for decades, until it was replaced by the formally constituted Diamond Trading Company, a wholly owned subsidiary of De Beers. For an incredible fifty years the Oppenheimer family held absolute sovereignty in the world of diamonds. For as long as production was concentrated in South Africa, Namibia and, later, Botswana, diamonds were the unchallenged fiefdom of De Beers, apparently impregnable.

JOHN RUDD

With great-grandfather Charles Rudd having served Cecil Rhodes with distinction, and grandfather and father both De Beers men, it was hardly surprising that John chose to make his career in the diamond empire. He too started by working in close association with the head of the group, in this instance Sir Ernest Oppenheimer, to whom he was a particularly effective personal assistant. After many years of distinguished management in the industrial diamonds division, John became editor-in-chief of *Indiaqua (Industrial Diamond Quarterly)*, serving the international industry and carrying articles of interest to all diamond geologists. Keenly awaited by subscribers, it usually carried photographs of scantily clad maidens who, though often irrelevant to the articles they graced, brightened an otherwise cheerless subject. Even if John did grow more eccentric with the passing years, his intellect and memory were second to none; he was a fine raconteur and a wonderful friend. He was 82 when he died in August 2009, full of years.

What undid the monopoly, starting in the late 1970s, was technology. As long as exploration had been carried out on the ground, using techniques that were tried and tested but shrouded in jealous secrecy, the world belonged to De Beers. But a new age was dawning, an age of easy intercontinental communication, of helicopters for hire, and of scientific discovery emanating from universities in the public domain. Fortress De Beers was about to be penetrated.

In 1991 diamonds were discovered in Canada – not just in the tantalising ones and twos that had peppered exploration results in the United States for decades, but in quantity. And not by De Beers. Suddenly Canada was where all the action was – until Australia entered the fray. A massive deposit of diamonds had been found in Western Australia by the new boy on the block, CRA, a subsidiary of the powerful Rio Tinto group. At the same time the Russians, long a relatively minor producer of diamonds from the Arctic Circle, were scaling up production. An empire that, until recently, had seemed invincible was showing widening cracks.

If technology has given us a level playing field, with new listings by diamond explorers a regular event on bourses from Toronto to Sydney, it has provided another, less manifest service that in its quiet way far outstrips the normalisation of an industry. It has given us a window into the workings of planet Earth that we would never have had without the diamond: a looking glass on the other side of which is a wonderland every bit as bizarre as the one Alice encountered. Except that this wonderland – the cradle of diamonds – is not the creation of a Victorian clergyman's imagination: it is as real as the galaxies we see in the night sky. As unreachable, certainly, and, until a few decades ago, as unknowable, yet it's close, at our feet, and, page by page, technology is opening it up to us, just as it shrinks the universe about us.

We shan't go underground yet, though. Let us rather look at a world map showing the distribution of diamond production. We find a pleasing, if rough, symmetry to the pattern, especially if in our mind's eye we bring South America back to Africa and, in a far less predictable move, India back against Kenya and Tanzania, where – improbably – it came from, as Gondwana broke up. Far up to the northwest is Canada; and to the southeast,

Rio Tinto's Argyle mine in Western Australia: a modern mega-mine.

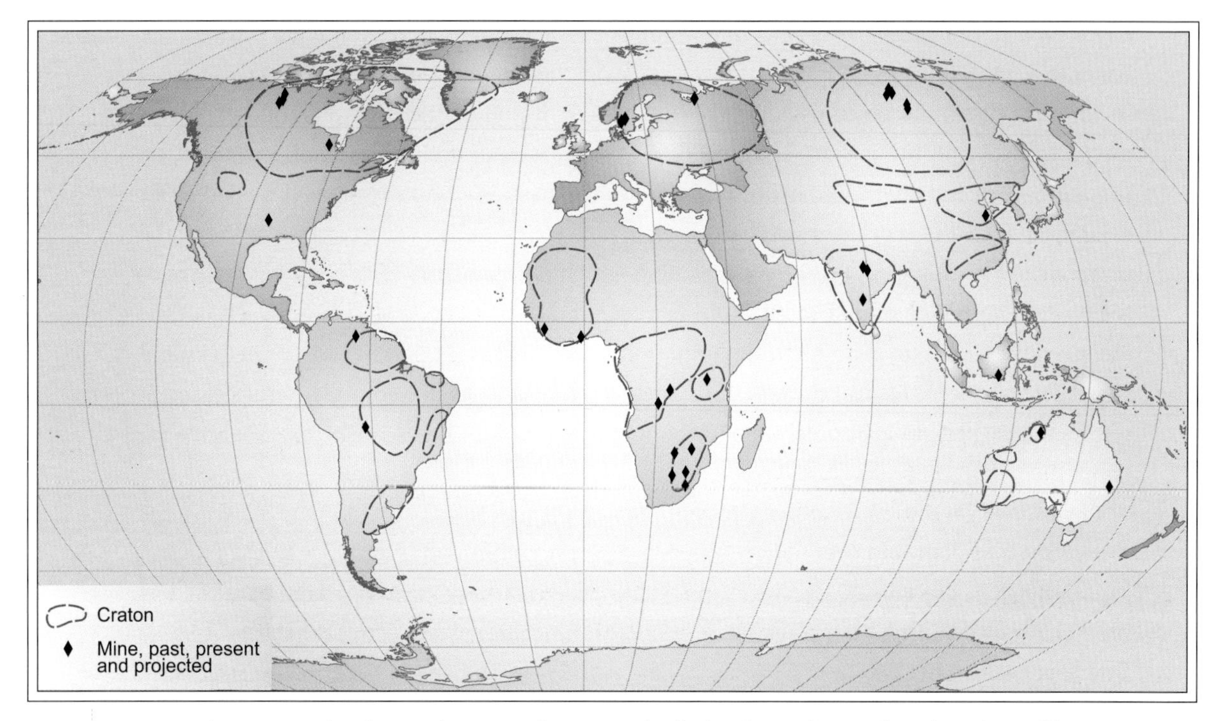

Approximate distribution of cratons and important kimberlite diamond mines throughout the world.

Australia. Where have all of history's big stones – gems so distinguished they have earned names for themselves – come from? India, Africa and Brazil. It's an approximate east–west axis. Where have the new finds of global importance been made? Australia, with a huge preponderance of small stones, none named, and Canada, with fabulously rich deposits of good stones though without the equal of Gondwana's giants.

But that's hardly science. Neither are the photographs we are shown of nodules studded with diamonds in a profusion that outranks any other concentration by orders of magnitude. Nodules are exotic, dense, beautifully travel-rounded fragments, mostly smaller than a rugby ball, some packed with garnets, and found only in a few select kimberlites. To those who first studied them – and all since then – their extraordinary mineral make-up and geochemistry point to an origin in the mantle and make them intensely worthy of research. Now we know that they and the diamonds were fellow travellers in the kimberlite.

This is new thinking, though. To earlier generations of geologists it was axiomatic that diamonds were an integral part of the kimberlite in which they were found. No one questioned it. Diamonds are unique, kimberlite is unique. Surely both formed in conditions like no other mineral, no other rock; in a layer of the Earth we call the asthenosphere, below the solid part of the mantle called the lithosphere and by far the deepest source of any magma we see crystallised at the surface today.

A cluster of diamonds in a mantle-derived eclogite nodule, the ultimate 'mother lode'.

Slowly a gathering body of evidence has opened another possibility, a new model that would demolish the old. Most of that evidence came from the nodules, and from the realisation that it was only where there was thick craton – as the most ancient continental nuclei are called – sending roots deep into the mantle that the kimberlites collected their precious load. And with this new premise came the realisation that the diamonds did not start the upward journey with the kimberlite magma; they were opportunistic hitchhikers, collected along the way.

That seminal breakthrough was made thanks to a few dedicated specialist geology professors and to deductive reasoning that would have made Sherlock Holmes or Hercule Poirot proud. And it is those rarest of rare diamond-bearing nodules – only a few have been found on six continents during centuries of mining – that are exhibit A, proving beyond any doubt the consanguinity, the shared genealogy, of the nodules and the diamonds.

But there's a deeper mystery than where the gems came from. Scientists have known for decades that diamonds are more than the hallmark of brilliance and hardness: each one is a miracle. They know that crystalline carbon – as diamond – is not stable at the ambient

Diamonds from the Ellendale mine in Western Australia, mostly highly resorbed.

temperature and pressure at Earth's surface. Diamonds shouldn't be here: their shiny crystals should have oxidised to common carbon dioxide, long before they reached the surface. Science says it's impossible, yet there they are, defiant in their perfection.

A decade or two ago we would have said the case was beyond solving. What we couldn't know then was that time would tell all; that we simply had to wait for technology to come of age. Then a steady stream of clues – from the kimberlite, from the nodules and from the diamonds themselves – would reward countless painstaking hours of research, and from the assembled clues would come the long-awaited answers.

Perfectly preserved diamonds from a Canadian kimberlite.

Now the history can be told, for science has given us, in a grand climax, the key to unlock the diamond itself and prise those last secrets from deep inside it. It is an unfolding story more dramatic than any fiction. We should be thankful for the diamond: brilliant emblem of endurance, essence of romance and, not least, messenger from a world more dramatic than even Jules Verne could imagine.

CHAPTER TWO

WHERE ON EARTH?

I watch a butterfly flap heavily past me in the clearing. Within a minute it's far down the stream, an occasional iridescent blue flash glinting in the sunlight. Beyond it there's movement and, as my focus changes, I see a man coming towards where we're sitting along the edge of the stream. He has a pack on his back with a big, round, coarse-mesh sieve fastened to it. 'Pork-knocker', one of the crew with me says: it's the word they use in Guyana for the fortune-hunters who work the river gravels and bank sediments for diamonds and gold. Not far to the south, in Brazil, they call them *garimpeiros*. We exchange news. We're clearing sites where in a few months' time drills will be probing the rocks at our feet for gold; he's after diamonds. 'Got a good one last year,' he tells me, 'seventeen carats.' I've often noticed the little shop fronts in the back streets of Georgetown advertising themselves as diamond dealers. I've never imagined that they might offer anything bigger than a carat, though. He smiles: perhaps my eyes widened.

Though not uncommon, a sighting of blue morpho butterflies in the diamond fields of Guyana and Venezuela never fails to delight.

Three months later, my term in the field done and the family reunited, we're celebrating Christmas with cousins in Oxfordshire. They've invited friends to join us for supper one evening while we're there, glad that we can meet because he too is an exploration geologist. He tells me about the company he heads, prospecting for diamonds in Liberia and Sierra Leone, both countries the darlings of diamond explorers now that they've been delisted as suppliers of 'blood diamonds'. He's optimistic: they've tied up some of the hottest property in that corner of Africa.

In my mind's eye I rotate South America a few degrees anti-clockwise and push it as close as it will go into the waiting embrace of Africa. It's a good fit. In the reconstruction Guyana, on the Caribbean coast of South America, and Liberia and Sierra Leone, on West Africa's south coast, are as close as neighbours on the same street. For hundreds of years

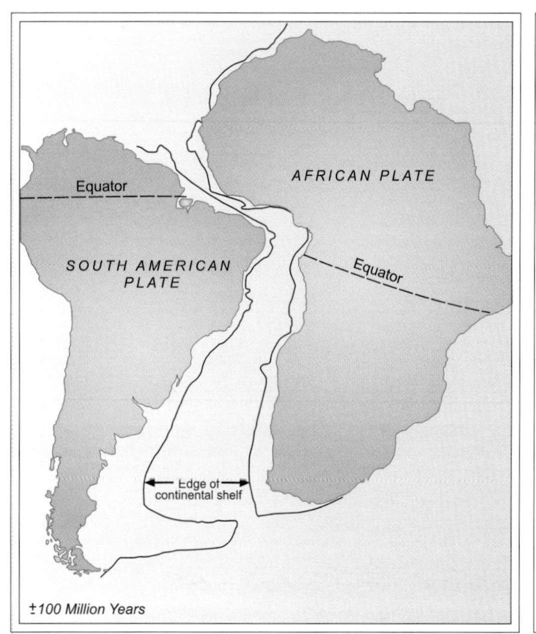

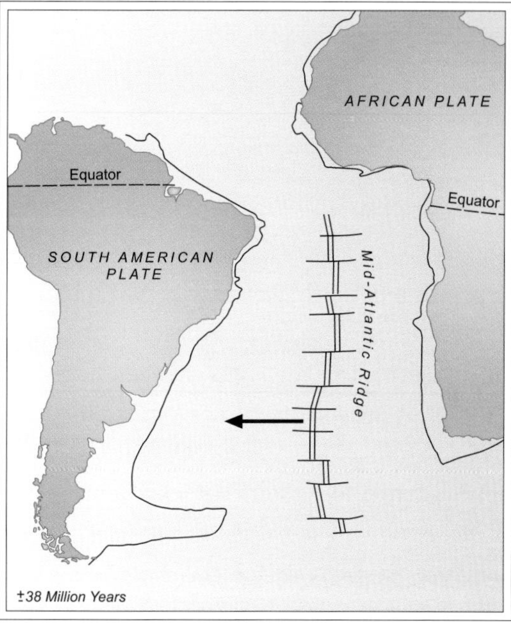

As the South Atlantic opened up, the drift was faster in the south to start with, and the northerly separation has still not quite caught up.

geographers have seen the fit I'm seeing. Some were so struck by the match of the opposed coastlines that they explored the possibility that the continents might one day have been joined; and not just those two. There was other evidence for earlier union, too, but inevitably it was circumstantial. Besides, the mechanism of how massive, deep-rooted fragments of *terra firma*, with their towering mountains, their rivers and lakes, their evolving flora and fauna, could glide over the surface of the Earth was beyond imagining.

But we need to leave this for a while and get back to diamonds. We'll see soon enough how the story of the architectural linkage – between plate tectonics, as continental drift has become known, and diamonds – unfolds.

Not long after the mechanism of plate tectonics was universally accepted, a geology professor at the University of the Witwatersrand, Tom Clifford, unlocked the secret of why pipes of diamond-bearing kimberlite – the mother lode – occur where they do. It was a sufficiently ground-breaking hypothesis that diamond geologists still refer to 'Clifford's rule'. Ignore it at your peril, they say.

It deals with a geological term that will reappear time and again in the pages that follow, so it is as well to explain it here. The term is 'craton'. It comes from the Greek word *kratos*, meaning strength, and is used to describe those parts of the Earth's continental crust that since very early times have remained rigid and stable – islands in a sea of more mobile crust. By definition cratons consist of very ancient rocks, mostly granite. They make up a

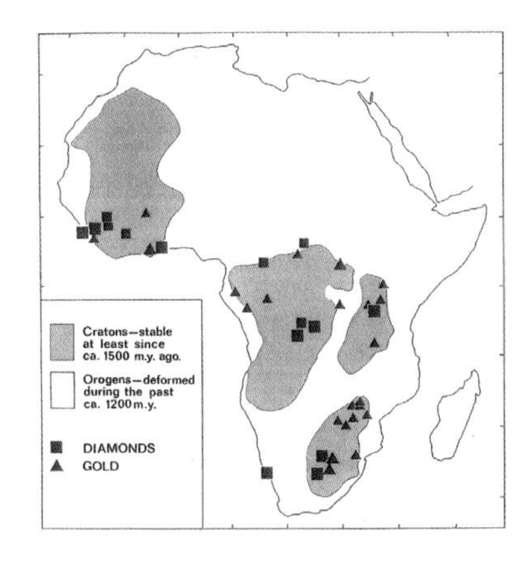

TOM CLIFFORD

Today Tom Clifford's map, shown here, looks unremarkable. In fact, it marks a Eureka moment in the history of diamonds. It was Tom who recognised in 1966 the fundamental subdivision of African geology into ancient cratons – regions that have been rigid for more than 1500 years – and, around them, post-cratonic fold belts; and it was he who first saw that different classes of mineral deposits generally show a distinct preference for one environment or another.

For nearly a hundred years before that, diamond explorers checked every kimberlite pipe with equal rigour and fervour. Now it is only the pipes within the ancient cratons that are followed up with the expectation that they may be diamond-bearing: those 'off-craton' must wait their turn with little hope of encouragement.

Forty years ago, the application of the concept of older cratons and younger orogens (or mobile belts) to the geology of Africa was innovative thinking. It was the product of work carried out under the aegis of the Research Institute of African Geology at the University of Leeds, initially funded, through the kind offices of Sir Ernest Oppenheimer, by the Anglo American Corporation, and directed by Professor W.Q. Kennedy and, reporting directly to him, Dr Tom Clifford.

Tom is a man of both humour and humility. He was as amused as he was amazed when he found, in the early 1990s, that a former research student had referred to the principle whereby diamond-bearing kimberlites occur almost exclusively within the cratons as Clifford's rule, a name first coined by Russian kimberlite geologists. Now diamond explorers from the Canadian Arctic to the jungle of Sierra Leone are familiar with the rule; they know that they ignore it at their peril.

With the advances made in the last four decades of geological research, we know a whole lot more today about the Earth's mantle which tells us why Clifford's rule works. We also know that it's a rule not a law, and that there are exceptions. As a basic axiom, though, it is the universal first principle of diamond exploration.

Gneiss of the Natal Metamorphic Province, not far south of Durban, typifies the sort of rocks that can be seen in mobile belts.

relatively small part of Earth's surface, a lot less than half of the continental crust. Compared with the terrain around them cratons are thick, deep-rooted in the mantle below them, and it is through them that the diamond-bearing kimberlites are intruded. In its simplest terms, that is Clifford's rule.

Around the cratons mobile belts are wrapped, cordons of folded rocks that, though they might have started life as sandstones, shales and basalts, have been cooked up beyond recognition during the folding, which was often accompanied by intrusion of hot molten granite.

Not long after diamonds were discovered along the Skeleton Coast of Namibia, a cluster of several dozen kimberlite pipes was discovered a short distance inland, near a village called Gibeon. This was long before Professor Clifford applied his thinking to the global distribution of these pipes, so singular in their mineral composition, and there was great excitement. Had these perhaps been the source of the fabulous deposits along the coast?

Pipe after pipe was thoroughly prospected. At first the negative results were a disappointment. But hope is in the prospector's blood. Gradually, though, the anticlimax that followed the excited anticipation of laboratory findings became weary resignation, acceptance that the family of pipes on which so much hope had been pinned was, after all, sterile. For the first hundred years of diamond exploration the experience would be repeated time and time again. But once Tom Clifford had published his observation that only 'on-craton' kimberlites were diamondiferous – and even of those, only a fraction – it did not take long for a new model to be embedded in the minds of diamond explorers. This was very simple: don't allow 'off-craton' kimberlites, such as the Gibeon field, to quicken your pulse.

A fifth 'C' had been added. For as long as diamonds have been traded, diamantaires have spoken of the 'four Cs' – criteria used by knowing buyers of polished gems to determine their value. They are carats, cut, colour and clarity, and it is they that will determine the price of a stone. But that's for guidance at the end of the diamond pipeline. At the beginning it's simpler, a single 'C': craton. Only one of the 'four Cs' used in valuing polished stones is straightforward and directly measurable: the size. The others require experience and intuition, which the best diamantaires have in abundance. For managers designing exploration programmes it's less

The Sand River gneiss is in the Musina Mobile Belt, where the important Venetia Mine of De Beers is located.

clear-cut, even if there is only a single criterion. No amount of talent can replace the hard yards, the thousands of man-hours that, slowly but surely, define the craton edges.

One of the problems is knowing what lies buried beneath younger, obscuring sediments, like the thick and hugely extensive Karoo Basin and the equally vast Kalahari Basin in southern Africa. Is it craton or mobile belt? Suddenly a whole new impetus was given to understanding the geology below the young sediments. Suddenly it was important to know what the basement was – and it's a question that tests geologists as much today as it did then, when the principle was still in its infancy.

For those not caught up in the explorers' race to find the next new fertile kimberlite pipe – for government geological surveys and universities – Clifford's thesis was a seminal discovery of equal importance, even if its application was more cerebral than practical. Geology had turned a corner: a whole rich, untapped field of research lay ahead.

Sometimes the flow of information and understanding was reversed. Instead of exploration companies taking their guidance from the universities, a discovery made by a company would lead to a radically new understanding of regional geology in the ivory towers. In the 1970s a De Beers exploration team found a major new kimberlite pipe, Venetia, in a geological environment which contemporary conventional wisdom had ruled out as non-prospective for diamonds. The find resulted in a complete overhaul of the interpretation of the strip of rocks known as the Musina Mobile Belt. Previously the belt had been looked at as a ribbon of younger, intensely deformed rocks separating two neighbouring but very distinct ancient cratons – and never the twain would meet. Now we know that they were joined, and the two former units can be seen as parts of the same craton, separated by a less stable but – and the distinction is important – not divisive strip of younger folded rocks, a relatively shallow feature.

In academia maps were reinterpreted and new models of crustal geology developed before the dust gradually settled. Although no one has said it until now, it was the unrelenting search for new diamonds that took the modelling of the southern African crust an important step forward. Since then even greater steps – giant strides in fact – have been taken, but they are due to geophysical and geological technology so sophisticated they were undreamt of when the Venetia deposit was discovered.

Though the jump in our story from drifting continents to the building blocks that make them may appear puzzling, there is a connection. All we have done is to cross an almost imperceptible divide between different layers of crustal architecture.

To explore this, let us stand at the world's highest kimberlite, Letšeng-la-Terai, in the Mountain Kingdom of Lesotho. We're at 3250 metres, the air is clear and the view over the stepped grasslands breathtaking. Letšeng is a singular mine in one respect other than its elevation: it is by far the lowest-grade kimberlite pipe mined anywhere in the world, yielding a niggardly 3 carats for every hundred tonnes mined. Think of it: a 3-carat diamond is the size of a pea and a hundred tonnes is far more than the load of the heaviest truck you'll see on the highway. To understand how the operation can be commercially successful, you have to know that all the money is in the super-big diamonds that are quite a regular occurrence in the Letšeng sorting house: the 50- to 100-carat stones and the occasional monster of 600 carats. So if truckload after truckload carries no more than a small diamond or two once a month, or every several months, one is carrying a gem that will be sold for millions or tens of millions of dollars.

Gem Diamonds' Letšeng-la-Terai, the highest kimberlite diamond mine in the world, with the lowest grade.

At Letšeng you probably didn't need your eyes to tell you how high you are: the chances are your lungs told you long ago. But why do we dwell on the height of Letšeng? It's by no means the highest mine in the world. The well-known silver mine at Potosí in Bolivia, now mined out, was over 4000 metres above sea level, and a number of currently active mines in the Andes are close to that. The Andes, though, and most other towering ranges are nothing like the Lesotho ranges, either in their general character or in their geology. They are all plate collision mountains, unremittingly

The gently rolling uplands of Lesotho are being gnawed away by rivers, slowly but surely.

rugged domains of steep slopes, sharp ridges and pointed peaks. The Lesotho mountains are manifestly different. Even as you look up at them from far below, the flat crest lines might give a clue as to what you would find if you were to negotiate the precipitous slopes they cap. You would be right if you had expected a surface flat enough to be easily traversed by any light-weight motor car, with goatherds watching their shaggy flocks roam the sparsely grassed plains – and, if you knew where to go, diamond mines.

These are not mountains pushed up by forces deep in the mantle; they are only as high as they are because their great antiquity has seen a hundred million years of soil erosion and stream and river transport reduce what was once a vast plain in the middle of Gondwana to a tiny remnant. Like an iceberg floating ever higher as it diminishes in size, Africa has risen. And if the flatness suggests stability, with no threat of the lethal lava flows or earthquakes that you might expect at this elevation on any other continent, it's because there's a great cake of stable craton below the goats and the diamond mines.

Let's pursue the antiquity of the Lesotho summit surface a little. Is it really a legacy of the plain that would have covered the whole of central Gondwana? The answer is 'not quite', because given that a hundred million years have passed, even from that relatively pristine central remnant some material has been stripped. What gives us an idea of just how much the surface has been lowered is the kimberlite itself. If there had been little or no removal of material we would see the crater of the volcano as it was when it erupted. But that's not what we see. We find that we're at least some tens, perhaps even hundreds, of metres below where the skyrocketing kimberlite burst through the surface, already down into the neck of the volcano. The crater and all the rock around it have been stripped off.

In many parts of the world where the kimberlite pipes were intruded into crust that was low-lying and not destined for great heights, the volcanic craters were covered over by deposits of sediment soon after eruption. There they have been preserved until today, only exposed again now that the covering layer has been skimmed off by natural processes. There are good examples of kimberlite-filled craters from the frozen north of Canada to the equatorial forests and savannahs of Africa. But not in Lesotho.

Having learned that diamond geology has given us the capacity to 'see' through kilometres of horizontal cover formations and know that the foundations of this part of the crust so deep below us are made of ancient stable craton, let us see what other revelations it has to offer. To do that, we'll choose a diamond and follow it – in a very telescoped way – along its course out of the heights and to the west coast of southern Africa.

The first stop we make is far from the mountain country at a farm called De Kalk, where the Orange River winds through the gently rolling interior plateau of the country. Our stone is about halfway on its 2000-kilometre epic journey. It started in a tributary of this river and will stay in it until it reaches the coast. In another 50 kilometres the Orange will be joined by the other main westward-flowing river in South Africa, the Vaal. Over the aeons both rivers and their tributaries, big and small, have brought untold fortunes of diamonds from their kimberlitic sources to the Atlantic coast.

We have stopped here for a variety of reasons, both geological and historical. The first noteworthy aspect of the geology is the cliff face of river-borne gravel perched high above the languid meanders of today's river. We see how in former times this same Orange River flowed at a much higher level than now, and how gradually, as it pursued its relentless incision, it has dropped and dropped. Riverside exposures of high gravel of this quality are extremely rare: geologists love them.

But they probably delight even more in the next bit of geology we encounter, still on De Kalk. Not far downstream from where we've been standing, and a hundred metres up the slope from the river, is a small exposure of rock that most people would walk past without a glance. Geologists, though, spot a glaciated pavement from far off, and stop in their tracks, pinching themselves to see that they're not dreaming. They'll walk around it, reluctant at first to set foot on it. If they have a camera to hand, they'll photograph it; if not, they'll just admire it.

In the Carboniferous Period (about 300 million years ago), vast areas of the northern supercontinent, Laurasia, from Pennsylvania to Poland to the Pechora Basin in eastern Russia, were covered by temperate forests, which would one day make the Northern Hemisphere coalfields. Far to the south ice sheets were grinding their way from the mountains towards the inland sea of Gondwana. As they inched forward, they caught up loose debris on the surface and incorporated it. And as they crept, centimetre by snail-paced centimetre, the fine dust they accumulated polished the rock they traversed, and the rock fragments scored it. They left behind smooth pavements, polished and scored, with the directions of the scoring – the striations, or striae – giving valuable information as to the direction of ice movement as, millions of years later, they became exposed. Considering that ice sheets, like glaciers and, for that matter, like rivers, obey the laws of gravity, the directions of the striae, coupled with the provenance of the material deposited as the ice melted, provide intriguing clues as to where the ice sheets had started; where the high ground of those times, so long ago, had been. So it is not just for the evocative image of ice sheets on a scale unknown today that we marvel at the striations on the polished surfaces. It is for the vital information they give geologists, the Hercule Poirots of earth science today.

The glacial scratches, or striae, on the De Kalk roche moutonnée *are clearly visible under the San graffiti.*

On the same polished, scored surface at De Kalk you'll see other markings, not just the straight lines that speak only to geologists. These are clearly recognisable as wild animals, drawn by human hands, not by a massive ice sheet. Archaeologists tell us that they were etched by early hunter-gatherers, who revered and drew the animals they hunted: the eland, the rhinoceros, the elephant. They celebrated their relationship with these creatures by immortalising them on the only surfaces that would preserve them for posterity, the rocks. Where there were caves, they painted; but here they had to find an art form that would withstand the summer storms – etching.

That these San (or Bushman) hunter-gatherers collected diamonds is quite probable. One of the respected pioneers of South African geology, E.J. Dunn, quoted a San tradition that they used small white stones from along the Vaal River to bore the holes in the stones that they used to add weight to their digging sticks. A century later, in 1970, a prospector on De Kalk came across what was undoubtedly a San tool-making 'factory', where he found a 3½-carat chip off a bigger diamond. Considering the area was away from any river-borne material, it had undoubtedly been carried there. The evidence for pre-colonial use of diamonds seems to stack up.

Whether the tradition continued into the time of the earliest 'Boer' or white settlers is open to speculation. It is a tempting connection to make, though, when one has seen the monument, a short distance away from the De Kalk farm house, that marks the spot where the 21¼-carat Eureka was found, in late 1866 or early 1867. As the name suggests, it was this stone that acquired fame as the first to be found in South Africa and would lead to the country's dominant position in the world diamond trade for the next hundred years.

From De Kalk we travel on, past the confluence of the Orange River with the Vaal, where our diamond and its fellows from the Lesotho highlands are joined by others from far to the north, and on. Finally we reach the coast. A strong longshore current carries our diamond northwards up the Skeleton Coast. A hundred kilometres north of the mouth it is caught in a local counter-current and settles into its boulder bed, its journey over.

For the last 400 kilometres of its journey in the river, and all the way up the coast, the diamond has been off the craton. The Kaapvaal Craton makes up roughly a third of the country, the northeastern third; to the west the Orange has crossed highly deformed formations where the crust is much thinner than further east. In the wide coastal strip between the Orange and the southern tip of the continent 700 kilometres to the south, there are kimberlite pipes by the dozen; and even though they don't flow annually any more, there are rivers that have carved out deep valleys. There are also diamonds along the coast near the river mouths, and some big mines. But making a connection between the kimberlite pipes, the rivers and the coastal diamonds would be to repeat an error that prospectors made until Professor Clifford's dramatic breakthrough with the craton connection. Every kimberlite

The scale of overburden stripping at this De Beers mine far south of the Orange River mouth gives an idea of the importance of the gravels below.

was rigorously tested, but not a single diamond did they yield. That's not where the coastal deposits came from. So where, then?

The Benguela current that bathes the African Atlantic coast with cold Antarctic water half-way to the Equator flows strongly and steadily northwards. It may reverse in local eddies, such as at the Orange mouth, but diamonds hundreds of kilometres south of the mouth were not brought by the Orange River that we know. In those words lies the answer. Research has shown that the course of the Orange we know is not the only one it has followed. The western half of the Orange has migrated, first south of its present position to only 200 kilometres north of Cape Town, and then north again. It does not take much crustal uparching of the flat, flat plains of western South Africa to deflect the course of the river, and that's what has happened. Tilting to the south and then to the north, the river has taken the line of least resistance and moved laterally, rather than cut down where it was. And wherever it debouched, it delivered diamonds, which in some cases settled quite far up its ancient valleys.

An even more intriguing case of diamonds far from where we might expect to find continental craton occurs in Southeast Asia. The curved belt incorporating Burma, Thailand, Sumatra and the island of Borneo is a part of the world that contrasts dramatically – in every way, not least geologically – with the archetype of stability that is the Kaapvaal Craton of South Africa. For instance, in 1883 this strip of crust gave us the catastrophic volcanic destruction of the island of Krakatoa, which left over 36 000 dead, and within the clear memory of all who read these lines, the 2004 tsunami, perhaps less devastating but still a

Bulk sampling at Gem Diamonds' Cempaka property in Kalimantan.

grim reminder of the restless planet we live on. There is a fascinating story that will be told much later in the book as to how the strange diamond deposits in this belt came to be. In the meantime, suffice it to say that the glacial striations we saw at De Kalk are not just a geological curiosity; they have, in fact, bizarre relevance to the Asian diamonds. Current thinking has it that at the time they were being formed, in another distant corner of Gondwana ice sheets were moving over a diamond-bearing craton that was later sliced off Gondwana. Like India it was plastered on to the Asian plate, though less forcefully, so that only part of the craton is attached to the mainland – Burma and Thailand – while the rest of it was added as islands – including Sumatra and Borneo.

These then are some of the stories diamonds tell us: about cratons where we thought there were none, about rivers that wander across the surface of the globe, and about extraordinary volcanoes that scavenge blocks of diamond-bearing mantle and hurl them to the Earth's surface. We might have discovered these things by other means at some time in the future, even without diamonds; or we might not. Remember, it is the inclusions in diamonds that give us data about temperatures and pressures of formation and about the age of the diamonds. Geologists hope and pray for stones with inclusions, whereas miners hope they will be perfectly clean and enormously valuable. Women – and some men – want pure diamonds, flawless and inclusion-free. Why, though, given that they are not as rare as sapphires or rubies or emeralds?

Let us just say that the reason lies somewhere within a combination of factors. Not only are they emissaries from the depths of the mantle and Earth's history and not only – on a finger, a throat or a breast – are they endowed with unique hardness, lustre and fire. Perhaps it is that they are in a class of their own, the only stone on Earth that can claim to be 'forever'.

CHAPTER THREE

FINDING AND MINING DIAMONDS

While there are still those who swear by a forked stick, big companies prefer greater sophistication.

How do geologists search for diamonds? Having found where they are, how do the miners and metallurgists pick those few carats from hundreds of tonnes of rock? And how do those ultimate craftsmen, the cutters and polishers, fashion dazzling gems from nature's simple crystals? In this chapter we take a glimpse through the window that opens on to those processes.

The year is 2006, the place Taung, South Africa, within sight of the location where the world's first 'missing link', *Australopithecus africanus*, was found 85 years ago. Three million years ago Taung Man would have hunted on these plains. The antelopes he sought would have gathered near the river at my side, today called the Harts River, and he would have stalked and trapped them here. He might not have gathered the pebbles and cobbles from the river banks, but later generations of African hunters would have used them to make arrow-heads and axe-heads.

I see those hard rounded stones through different eyes. To geologists the beds of gravel that flank the Harts tell of water that flowed powerfully enough to pick up rocks as big as a watermelon and carry them tens – some even hundreds – of kilometres from their source to this place. In the headwaters where these boulders started their journey were pipes of the kimberlite rock that had brought diamonds to the surface. Slowly the diamonds were fed into the tributaries of the Harts and carried with the floods that brought their rocky load to where I stand. Over millions of years the river gradients, together with the rest of the landscape, flattened. The climate became more arid and the floods less frequent. Now the rare torrents bring sand and mud, the stones that are yesterday's legacy barely jostled along.

I crisscross the area of old workings, where a few decades ago diggers scraped away the soil to expose the ancient river gravels that held the treasures they dreamt of, now far from the tree-lined bank. In one of the scrapings I find what I'm looking for: a vertical section with gravel exposed. I photograph it and mark it on the map I'm making, noting that there is no sign of anyone having been here for years.

The next day I'm back, now searching for what I'm very sure I won't find: illegal miners. My informant leads me through the trees and, amazed, I see a wisp of smoke curling out of the ground ahead. It disappears, but we have noted where it comes from and head for the place.

Looking at a stone on a finger, one might not imagine how much earth is moved to get them.

In a shallow pit two men stand as we approach, their half-eaten breakfast between them. Greetings are exchanged as we clamber down into the pit. Next to them is a handmade shaft just big enough for a man to lower himself down, a bucket with a rope tied to the handle near its edge. There's an old sieve and heaps of fine sand; nearby are piles of coarse material, from a rice grain in size up to a pigeon's egg. All the bigger stones – pebbles, cobbles and small boulders – lie in a corner of the pit. The coarse and the fine parts of the material are of no interest: it's the inbetween-size fraction that will carry the one stone that may be their passport to a life of prosperity and ease.

When they see we mean them no harm, they open up, and we talk. Their last stone, of just over 3 carats, they tell us, was found four months ago, and fifteen months before that they found a 7-carat stone of excellent colour. These are the objects that have them up at dawn every day, to tunnel for twelve hours along the base of the gravel, five metres below the surface, the danger of a fatal collapse of overlying rock always present. They are worth risking lives for.

A hundred kilometres away, next to the Vaal River at Windsorton, 3- and 7-carat stones are quite a regular occurrence for Louis Oosthuizen, who conducts a multimillion-rand operation there recovering diamonds from gravel. Let's imagine he flies in as we stand at the edge of a gaping pit into which you could fit a small village, his helicopter first heard when it is just about on us, above the roar of trucks leaving and re-entering the pit and the massive shovels clattering and screeching as their fearsomely toothed buckets scrape up great bites of gravel. The thud of chopper blades slowing at last, Louis shouts a greeting to us and walks to the edge of the pit. There he surveys it from end to end and top to bottom. The two minutes seem like ten. Satisfied that all is as it should be, he invites us to join him in a tour of the plant.

We see how all oversize material is screened off and the fines washed out, just as our two diggers had done, but now on a massive scale. We see various stages of gravity separation, as the lighter material leaves the trail through the plant. Next the concentrate – by now of a carefully limited size and density – goes through the X-ray machine, where all non-fluorescent pebbles leave the circuit temporarily, to rejoin it later. Most diamonds are fluorescent, our host tells us, and he needs to reduce the concentrate as far as possible before the train goes onto a belt where it will be visually sorted. To cater for the rare non-fluorescent diamonds, the rejects from the X-ray chamber will go directly to the grease belt. He'll take us there in a minute, but first we're going to see the visual sorters in action.

On the mine diamonds are sorted and counted in the 'glove-box'.

After signing in, we pass through two heavy iron doors, each double-locked, one key with one of the sorters, the other with a security guard. Shielding the entry code from us with his body, the security guard enters the numbers into a keypad next to the door. He has deactivated the tear-gas gun just inside the door of the inner sanctum. Now within the mine's own Fort Knox, we see the concentrate from the X-ray spilling through a chute onto a slowly moving belt in a sealed Perspex-encased chamber. It has four gloves built into it from one side, into which the sorters have inserted their hands. In the sorting chamber are tweezers, sealable bags, a scale, labels, a pen and camera. With their tweezers the sorters spread the train of concentrate over the width of the belt, scrutinising every stone on it. One of them spots a clear stone. He picks it up with his tweezers, looks at it carefully through his magnifying glasses and puts it back on the belt. It is not a diamond. It joins the rest of the concentrate, which, as the belt rolls under itself to return to the beginning, drops through a chute into a heavy drum below the chamber at its end. When full, this will be locked with a padlock and in due course it will also go to the grease belt.

Louis tells us that it may be a while before they spot a diamond, and we move on. Back in the bright Northern Cape sunlight we blink, our lightning tour over. We're not going to see the grease belt after all, but we do know that this is one of the oldest ways of capturing diamonds, using the property whereby, unlike just about every other substance, diamonds repel water. They do not 'wet' but stay dry when water is sprayed on them. This means that they stick to the grease when they tumble on to the belt, unlike the other grains that slip off because they are surrounded by a layer of water. After the X-ray and the visual sorting, this is the safety net which catches the last would-be fugitives.

It's a far cry both in scale and in technology from the two men standing over their breakfast fire at Taung. We've looked at two extremes. Between them are hundreds of other operations, and they're not very different whether you're on the plains of South Africa or in the Brazilian rainforest. A moment's thought, though, will tell you that even the 'high-tech' operation at Windsorton uses remarkably unsophisticated principles, ironically resonant of the simplicity of the stone that is at the centre of it all. Very little science is used to concentrate diamonds in recovery plants, compared with the sophisticated metallurgical processes used for extracting gold or copper or aluminium from their ores.

And it's not very different in nature. Unlike gold, many times denser than anything else in the river, diamonds – only a third as heavy again as Earth's commonest mineral, quartz – do not concentrate very effectively by virtue of their density. The diamond's most exceptional attribute is its hardness, but this, too, does not set it apart from other river travellers in a way that might help to harvest it, unless one were in a survival race to the end of time. Copper and aluminium – and just about every other metal you can think of – follow varied and elaborate processes in the Earth's crust to gather in the concentrations where we can profitably recover them. Not alluvial diamonds, though. If they weren't so singular and so sought after, we would leave them where they are, just as we don't try to recover copper from the soil in our gardens, even though it is there in the same concentrations as diamonds in river gravels.

There is a stretch of the Orange River that meanders across the dusty African plains between the towns of Douglas and Prieska where the average value of the diamonds is over US$1500 per carat, just as they are, in their bed of cobbles and boulders. The so-called Middle Orange River is the diamond field with the highest-value stones anywhere in the world. This makes it possible to mine grades as low as half a carat per hundred tonnes in some mines,

A big Orange River digging in the evening.

or one part per billion. All you've got to do is move 100 tons of gravel, or three truck-loads – and get the diamonds out of it – for less than $500. It can be done.

We need to emphasise that everything we've just said applies to river diamonds. If it's an industry booming today like never before – and it is – it's far from young and by no means restricted to a few rivers in South Africa. Alluvial diamond mines have been opened up over hundreds of years in India, in central and northern South America, and over a huge swathe of southern and West Africa. They may be one- or two-man hand diggings or they may be highly mechanised mines that have cost millions of dollars to establish. But the target is the same: the brilliant crystals that are trapped behind boulders or in holes in the floor of ancient river channels.

Throughout history, every diamond in both India and Brazil that has reached the market was won from alluvial diggings. It is probably true to say that the first discovery of diamonds in these countries was opportunistic. A child playing by the river, or a prospector panning for gold, spots a shiny stone. The stone is tested and found to be a diamond. Word gets out and prospectors flock to the place, spreading out like a wind-driven veld fire. Slowly a new field is defined.

In South Africa, and in every other diamond-producing country in Africa, the primary source that fed the stones into the streams and rivers, the kimberlite pipes, have been discovered not long after the alluvials. Many of the pipes became the basis for mines – some of them mega-mines – themselves. Africa seems unique in having a balance of alluvial and kimberlite diamond deposits. All the Indian and South American mines have been alluvial, and those in Canada, Russia and Australia are almost entirely kimberlite-based. Africa, with both, has played true to its central geography.

In West Africa not much has changed in a hundred years.

If the first finds of alluvial diamonds were serendipitous, how have they been followed up? To borrow a phrase, it has been more by perspiration than inspiration; more by trial and error than by the clever science used to find kimberlite pipes.

While in the West African alluvial diamond fields a high proportion of stones are found by divers groping in the muddy waters of today's rivers, South Africa is different. The ancient 'palaeochannels' that yield their bright bounty may in fact be out of sight of the river that formed them. And 'ancient' is a word not used lightly by geologists: the channels may be many millions of years old, and during that time the river may have meandered twenty kilometres from its earliest course to where we see it flowing today. Since its earliest days sand has blown in and indiscriminately covered everything on the flat surface it found, bedrock and river gravel alike. If you walk over the ground or fly over it, there is no sign of where the ancient channels were.

Once you have an indication that they are present somewhere, there is only one sure way to locate the palaeochannels and that is by drilling the ground. It is not deep drilling and the geologist does not need a solid core of rock: chips will do, so the cost is low and the work quick. The geologist has to be canny, though, to get the maximum amount of information from the chips blasted out of the ground – fragments of rock smaller than a man's thumbnail. He needs to try to decide whether they've come from boulders (ranging from very large down to the size of a honeydew melon) or cobbles (smaller, down to a lemon in size) or pebbles (anything else bigger than a grain of rice). In a nutshell, the bigger the better: big river rocks tell of strong flow, a wide catchment area sampled by the river and its tributaries, and good potential for trapping the diamonds it has collected. With experience the geologist can gauge quite accurately from those dusty rock flakes what he would see if it were all

With boulders weighing up to several tonnes, it is small wonder that big samples must be taken.

opened up with the big excavation necessary to take a sample and test its diamond content.

To evaluate the ground the miner needs a sample big enough to tell him not just whether he's got 1 carat per hundred tonnes, or 5, or none, but also whether he's got big stones or small, and are they worth $50 a carat or $1000? He needs to see a few hundred carats before he can decide whether to prepare his presentation to the bank manager or not. One consolation is that, unlike just about any other kind of exploration, the miner can sell his sample. If he's lucky he may even be able to make his prospecting pay for itself, and that's not something that happens every day in mineral exploration.

The massive surface installations at Cullinan Mine.

Small wonder, then, that there are hundreds and hundreds of small-scale diamond miners in South Africa and Brazil. But what of the hard-rock diamond mines, where the mother lode, the 'blue ground' or kimberlite, is mined? Are there small operators trying their luck there? The answer is a simple no. Crushing rock to liberate the diamonds you hope are there, and being sure your diamonds aren't themselves being crushed, is a costly business. And the only way of understanding whether the kimberlite pipe maintains its surface proportions below the ground surface is to core-drill it, a slower and much more expensive process than the drilling used in alluvial exploration. Because the stones have not been subjected to the upgrading that rivers accomplish as they smash the fractured stones and release the smaller stones into the fluvial mêlée, they are almost invariably of a much poorer quality than those from alluvial diggings. This means the grade has to be higher, generally as much as ten times.

On the other hand the mines are generally bigger, going down hundreds of metres into the Earth. The Cullinan Mine east of Pretoria, for example, has been open for just over a hundred years and is still a major mine by any standards; and some of the Kimberley mines continued for as long. With a longevity that very few metal mines can boast, it is no wonder that kimberlite pipes are the first prize for diamond miners. They are no easier to find than any other mineral deposit, but this has never deterred companies from spending annual budgets of millions of dollars searching for that one pipe in a hundred that carries 10 carats per hundred tonnes and – what they all dream of – the occasional giant stone of the finest 'water' (as the combined clarity and lustre are known), 'D' colour (as the experts term the best possible colour) and flawless.

Let us step aside for a moment and remind ourselves where the name for the mother lode came from. It's a story that is both strangely circular and ironical. As we have seen, in 1871 the first 'dry diggings', far from any river, were the discovery site of an equally new phenomenon, 'blue ground', a dark grey-blue rock like nothing anyone had seen before. Once the first geologists to see blue ground recognised it as something hitherto unknown, it had to be named. What also needed an identity at that time was the little settlement sprawled

Modern Kimberley, with the De Beers Mine in the foreground.

At De Beers' Jwaneng Mine in Botswana many metres of sand cover the bedrock.

around the 'dry diggings'. There was no debate about the settlement's name. As so often happened in the colonies, it honoured an important person back in Britain: the Secretary of State for the Colonies, the Earl of Kimberley. For those charged with finding a formal tag for the 'blue ground', the task was as straightforward. With no distinguished discoverer to pay homage to, 'kimberlite' was as good a name as any for the rock on which the town was based.

No one could have dreamt that Kimberley would become the diamond capital of the world. And it is in that later turn of events that the irony lies. With four major mines in its municipal boundaries – the Kimberley Mine, the De Beers Mine, Dutoitspan and Bultfontein – and with the spiritual, if not administrative, headquarters of De Beers Consolidated Mines at its heart, where monthly board meetings are still held, Kimberley's claim as the centre of the diamond world is secure. History has shown us that never was a rock more aptly named.

This digression is not as gratuitous as it might seem. With diggers seeking their fortune in the nineteenth century from Bendigo to California, whether from gold or diamonds, there was good reason why it was in the plains that sweep across the centre of South Africa that the mother lode for the river diamonds was first discovered. Not only were the diamonds large and plentiful, but they were near the surface in an area of sparse vegetation. The combination of these four factors is unique. In India and South America hundreds of years of exploration have shown us that primary sources for diamonds may not have survived; in Australia they do exist but are fiendishly difficult to find; in Canada and Russia they are mostly covered by

NIKOLAI (NICK) SOBOLEV

In 1941 Nick Sobolev's father, Vladimir, identified the Siberian platform, the Vilyuy region in particular, as most prospective for diamond-bearing kimberlites, on the basis of its similarity to the part of South Africa where most of the world's diamonds were being produced. The idea grew out of fieldwork he had conducted in Siberia in 1936. Nick himself is today considered the father of Russian diamond geology and is also remembered as the geologist who, far behind the Iron Curtain, pioneered the science of geochemically identifying diamond-bearing kimberlites. With only a tiny percentage of kimberlite pipes carrying commercial grades of diamonds, a quick, affordable technique that can sort the wheat from the chaff became critical, and in Russia it was Nick Sobolev who found the vital key. At 75 he still researches tirelessly and is a credit to his giant of a father. So, too, are his younger brothers, the twins Alex and Steve, both geologists, and Eugene, who distinguished himself as a physicist in the field of diamonds and sadly died in 1994.

glacial till; and in the rest of Africa they are covered either by the thick sand of the Kalahari Desert or by tropical rainforest.

Every mineral sought by mankind has its pathfinders, essential clues which allow geologists to pick up, and then follow, their trail. Diamonds from kimberlites are no different. Kimberlite rock itself has a distinctive mineral make-up, with some quite unusual minerals that occur in far higher concentration than diamond and decompose slowly enough to survive for a long time in the weathering and dispersal chain. Systematic sampling of an area with indications of kimberlite pipes will generally find them. The pipes are commonly different enough from the ground into which they are intruded to show themselves, either as geophysical 'anomalies' or in subtle changes of vegetation. The anomalies are picked up by regional surveys, usually aerial, measuring the magnetic or electromagnetic or gravity field – or all three – of the region. In other words, the pipes are more, or less, magnetic or conductive or dense than the rocks around them, depending on what these 'host' formations are. And any vegetational anomaly that is more or less circular and measures anything from a half-hectare to a hundred hectares is followed up, particularly if it coincides with a geophysical anomaly. Quite commonly, too, kimberlites stand slightly higher or lower than the surrounding terrain, giving geologists another clue to work with.

Detailed surveys of anomalies identified from the air define them better and may substantiate them or reduce their credibility. Sampling of soil and rock follows, in pursuit of the diagnostic minerals: particular garnets, shiny black ilmenite or the bright green chrome diopside. Then comes the acid test of drilling, to prove that the body giving rise to the indications is kimberlite and, supposing it is, to find how big it is.

Panning in the DRC for the indicator minerals that will lead to the kimberlite.

But kimberlite rock *per se* is neither particularly rare nor especially valuable. So the next question the geologist asks is whether the pipe is diamondiferous. Here's where the clever science comes in. Considering that the pipe may be in the middle of the Kalahari Desert or the remotest Canadian tundra, the obvious way of answering the question, by testing thousands of tonnes of rock, may not be an option. Thanks to thousands of man-hours of research around the world, there's a shortcut that is almost infallible: test the garnets and the other KIMs (kimberlite indicator minerals) for their chemistry.

Like the diamonds, the KIMs are not part of the kimberlite, although they would not have got to the surface without the kimberlite. The garnets and the others we spoke of earlier were plucked from deep in the mantle by the kimberlite in its ascent, as – most important of all – were the diamonds. The chemistry of those minerals gives geologists a wealth of information about the temperature and pressure of the very particular piece of mantle that was sampled. Unless these conditions were exactly right, diamonds would not have formed there, would not have been sampled by the ascending kimberlite. It may seem laughably simple, but consider for a moment the technology that has enabled us firstly to make the correlation between the chemistry of small mineral grains and conditions hundreds of kilometres below the surface of the Earth, and then to measure with great accuracy the chemistry of grains that may be no bigger than a pinhead. For now let it suffice to know that the magic phrase for kimberlite geologists is 'G10 garnet'. If the laboratory announces that it has found a high concentration of G10 garnets, it's time to break open the bubbly.

The Diamond Bourse in Antwerp.

◆◆◆◆◆

HAVING ESTABLISHED how geologists search for diamonds and how the miners and metallurgists extract them from the Earth, let us now go straight to the main part of that still unmapped stretch of the technical narrative: the production of the finished gems.

Like the journey so far, the cutting and polishing of rough diamonds takes us around the world, out of the wild places to Antwerp, Amsterdam, London, Tel Aviv, New York, Johannesburg and Surat in India. These are the centres of gem production, or diamond manufacture as it is more correctly called.

Just as technology has made giant strides in the exploration for diamonds, it has played an important role in the unexpectedly complex world of their cutting and polishing. Perhaps the complexity of it ought not to surprise us. Here we are on the brink of the realm of mystery. We are, after all, dealing with a gem whose hold over us is as powerful as it is unfathomable, yet it is fashioned from the simplest crystal Earth has to offer, one whose value soars highest when absolutely devoid of any hint of colour. With none of the green of the emerald or the red of a ruby to catch the eye, the diamond depends for its magic on its extraordinary capacity to concentrate brightness.

Not only does a well-cut gem catch the light and return it to the eye, it breaks the white light that enters the stone into its spectral colours, so one sees flashes of yellow and blue and red. This simply does not happen with other stones. The secret lies in the diamond's high refractive index, or RI. This is the measure of the extent to which light is 'bent' on passing from air into another medium, in this case the diamond. To give an idea of what makes the diamond special, the RI of water is 1.335, of quartz 1.547, of emerald 1.582, of sapphire 1.765, and of diamond 2.419. What these figures indicate is that the diamond has by far the highest RI of any gemstone. Jewellers talk of the 'fire' of a gem, meaning the show of spectral colours as it refracts white light before returning it to the eye. The diamond's unique fire is the direct consequence of the stone's high refractive index.

Light emerging from a diamond is broken into the colours of the spectrum.

In return for the fire the diamond gives us, we reciprocate by giving it 'life'. This the manufacturer does by the way he shapes the stone so as to maximise the reflection of the light that enters the stone. His skill lies in designing, then creating, the 'cut'. The evolution of the different cuts applied to diamonds, as with other gemstones, has relied more on geometry than technology, with the two elements, fire and life, balanced against each other in a complex array of reflecting surfaces – the facets – for optimum effect.

The Brilliant Cut, especially the Ideal version of it, as it is formally known, is widely regarded as being the optimum for displaying the 'brilliance' – the combination of fire and life – of the stone. But this may not suit the particular taste of the person who wears the stone. Added to the question of personal fancy is the fact that the Ideal Cut was only designed in 1919, when dozens of famous stones had already been cut. So, over time and according to differing opinions among designers as to the optimal balance of fire and light, dozens of different cuts have evolved. The accompanying figure shows the Oval Cut, the Pear-shaped Brilliant, the Marquise, the Heart-shaped Brilliant, the Baguette, the Emerald and the Square Emerald. These are but a few of the distinctive and better-known cuts.

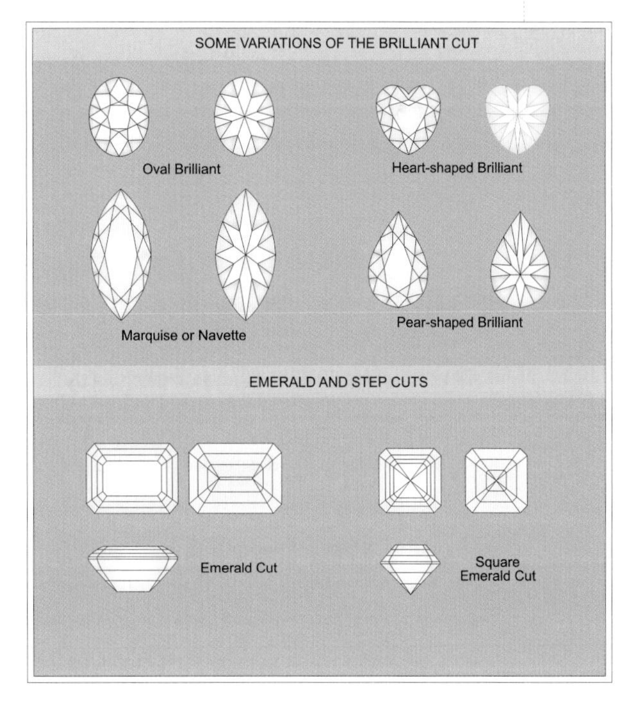

Maximising the brilliance depends on balancing the angles and the number of facets, aspects which have taxed the minds of

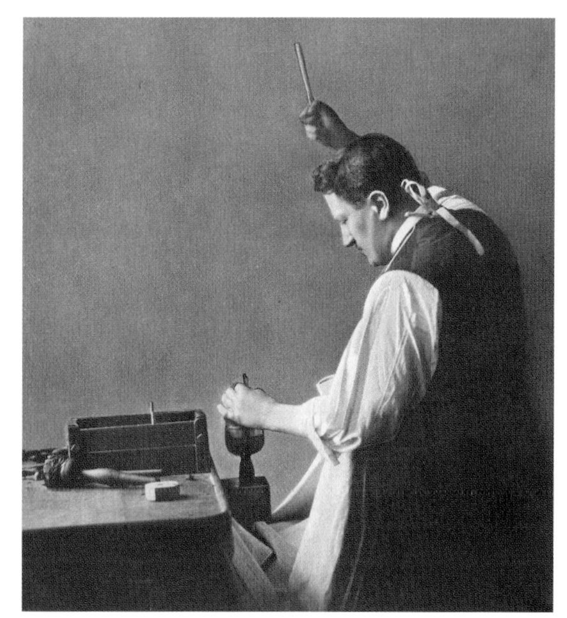

Joseph Asscher marking and then cleaving the Cullinan diamond.

designers for centuries. The Ideal Cut was the product of a young mathematician, Marcel Tolkowsky, who earned a master's thesis for his work, and immortality. He designed, on theoretical grounds, the configuration of facets that would give the greatest fire for the least loss of life. He is said to have started his work by canvassing members of the public as well as cutters in the family business in Antwerp as to their preferred gem of a small range of different cuts, and then applied his mathematical skills to perfecting the cut. Whether or not this story is apocryphal is irrelevant. What is germane is that his cut was subsequently slightly modified to provide a girdle, for greater strength of the mounted stone, for his cut had, in effect, no girdle – what is termed a 'knife-edge' girdle in the industry.

The main consequences of using angles in the cut other than those of the Ideal Cut are that light escapes through the pavilion, or back or bottom of the stone, and that in seeking to minimise this leakage of life, fire is lost. The ideal is a delicate balance, requiring consummate skill by both designer and cutter.

So much for the wonderful world of diamond manufacture – part science and part art. Where are the pitfalls, and how far have advances in manufacturing kept pace with those at the front end of the industry?

If the difference in weight between a 0.95-carat cut diamond and a 1.05-carat stone is 10 per cent, the disparity in value is far greater, as much as 30 per cent. In the case of marginal 1-carat cut stones unscrupulous cutters will 'cheat' by making a 'spread' cut, knowing that the average eye will never discern that slight loss of brilliance. Just as certain is that the jeweller's scale does not lie. The deception happens, though no one knows how much. *Caveat emptor*: buy from a reputable dealer.

Faced with a particular rough stone, the designer has to decide how he is going to cut it for optimum value, without sacrificing brilliance. If he values his reputation he will want to ensure he finishes with a good, or fine, 'make'. Lack of scruples or inexperience will give him a make that is described as poor or bad. But before he decides how he will cut the stone and what sort of weight gem he can produce from it, he will study the form of the rough diamond. It hardly needs to be said that the bigger the rough stone, the more important his decision. Can he get two – or, if it's a really big stone, more – gems out of it? The shape of the rough, too, will decide whether he goes for a Brilliant or one of the other cuts.

How two Brilliant Cut gems might be cut from a single octahedral rough diamond for maximum yield.

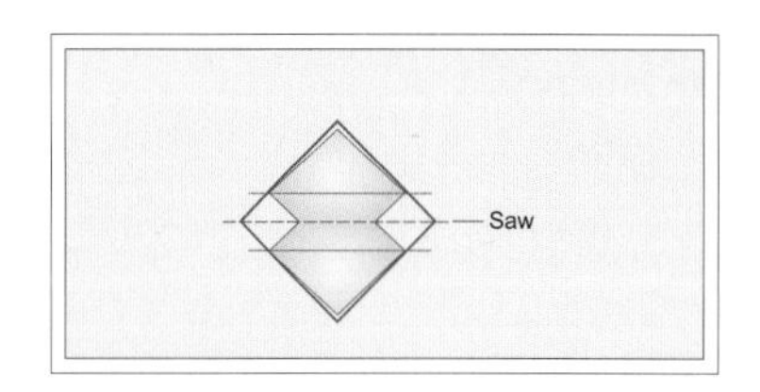

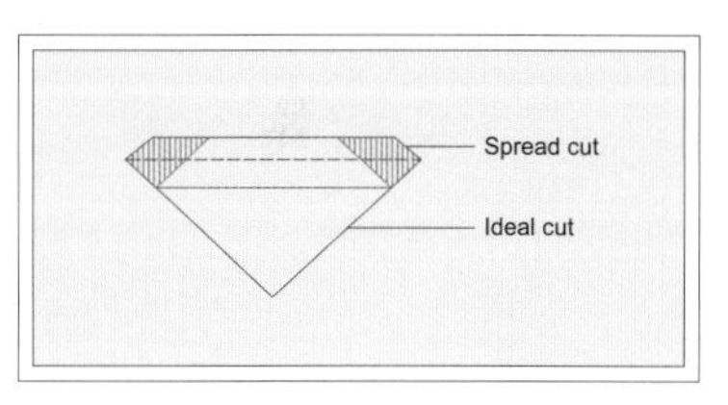

How unscrupulous manufacturers cheat to get the magic 1-carat gem from a rough diamond that doesn't honestly have it.

Before the technological revolution, the designer clearly had to be an artist. Now scientific help is at hand. Just as computers have made anyone who can afford a quality digital camera a photographer of near-professional calibre, they have burst into the field of diamond manufacture. Put a rough diamond into a computerised instrument and it will show you your various options of cutting. Now, instead of using a fine diamond-studded saw to cut your stone for its basic size and shape, a laser will do it for you, with greater precision. The final shaping and polishing, though, is still the preserve of skilled professionals with sharp eyes and steady hands. All day long they bend over spinning laps, shaping facets so small they can barely be seen with the naked eye, perfecting the angles between facets. Seeing it is to go through the looking glass into a fantasy world.

We shall now leave the high-tech world of modern diamond manufacture to explore how it was done in a more gracious age, and how it has evolved over the centuries.

The Indian diamond capital Golconda in its glorious heyday.

CHAPTER FOUR

INDIA AND SOUTHEAST ASIA

Around 600 BC the Hebrew prophet Jeremiah wrote: 'The sin of Judah is written with a pen of iron, and with the point of a diamond ...' No one can say how long diamonds had been known before then: all we can be sure of is that the diamonds of Jeremiah's time were from India, the cradle of diamond mining. For over two millennia, India was the only supplier of diamonds to European monarchs and Persian potentates. And if for nearly three hundred years it has occupied steadily descending rungs on the ladder of the world's producers, it has recently reinstated its pre-eminence in another way. India now cuts nearly three-quarters by weight of the total global diamond production. By value it is less than this because its manufacturers specialise in stones that are too small to be profitably cut and polished elsewhere, and one of the basic facts of the diamond industry is that size and value are intimately related. All other things being equal, a 10-carat stone is worth ten thousand times more than one of a tenth of a carat.

India did not, however, make its way into the history books with small diamonds but with single large gems that would proudly adorn crowns and turbans and other royal regalia. Yet in the dawn of diamond days their value was hardly more than that of gold, and that only because of magical or mystical properties. Today top gems are worth ten or a hundred times more.

Gold has always been valuable, even if only more recently as a basis for currency and a measure of national or personal wealth. In earliest times its value was for the way it could be used for personal adornment and in decorating the finest palaces and temples. Idols were made from gold, and their hallowed sanctuaries were richly gilded. The Lord's instruction to Moses for the Ark of the Covenant specified the liberal use of gold, the wooden chest of the ark (about a metre and a half long, and just under a metre high and wide) to be overlaid inside and out with it, as well as gold mouldings, gold carrying-rings and hammered gold cherubim.

The value of gold lay not in its easy availability; far from it. Copper and tin were obtainable in much greater quantities, but first they had to be refined from the ores in which they were found, whereas on the surface gold occurred in its 'free' state naturally. What is more, the earliest metalworkers had found that while copper and its alloys tarnished over time, gold retained its rich colour and incomparable lustre year after year, century after century.

How appropriate that gold would lead the way to diamonds. Thousands of years ago the first Indian diamonds were recovered in pans where sand and fine gravel were being washed for the gold that was so eagerly sought. Mostly the little stones were frosted and had no particular form. Recovery of the occasional lustrous stone was cause for celebration, particularly if it had the octahedral shape that showed up from time to time. They were impossible to cut, though. Small wonder that the first use of diamonds for personal ornamentation is not well documented.

What is better known but harder to understand is the value diamonds had from earliest times for reasons mythical, mystical and magical. Even more curious is that from early times they found a market with the Romans, who we are told also used them as talismans.

One of the most valuable references for students of early Indian history is the *Artha-Sastra* of Kautilya, written in the fourth century BC. Kautilya was a minister and main adviser in statecraft to King Chandragupta Maurya, who, together with his son, Bindusara, consolidated for the first time most of what we know as India today, in the late fourth and early third centuries BC. Only the southern tip and the island of Sri Lanka did not fall under the dominion of the Maurya Empire. Kautilya's *Artha-Sastra* is probably of no less significance in Indian history than the Magna Carta to the English, and is made much of here for its reference to *vajra*, or diamond, the first formal and reliable mention known.

The *Artha-Sastra* proclaimed that a diamond that was 'big, heavy, capable of bearing blows, with symmetrical points, capable of scratching a (glass) vessel, revolving like a spindle and brilliantly shining is excellent. That with points lost, without edges and defective on one side is bad.'[1] The rare octahedron, with its exceptional and mystically inspired radiance, was the ultimate ideal.

The next comprehensive description of the value of diamonds comes about 800 years later, from the *Ratna Pariska* of Buddha Bhatta, which dates from late in the sixth century AD. *Ratna Pariska* can be translated as *The Estimation and Valuation of Precious Stones*. The rainbow reflection of clear white stones was the key to their value, and octahedra that displayed this were to be reserved for kings. The mystical powers of the diamond were as revered as they had been from the beginning, a flawless octahedron guaranteeing to the wearer 'happiness, prosperity, children, riches, grain, cows and meat', and to kings specifically

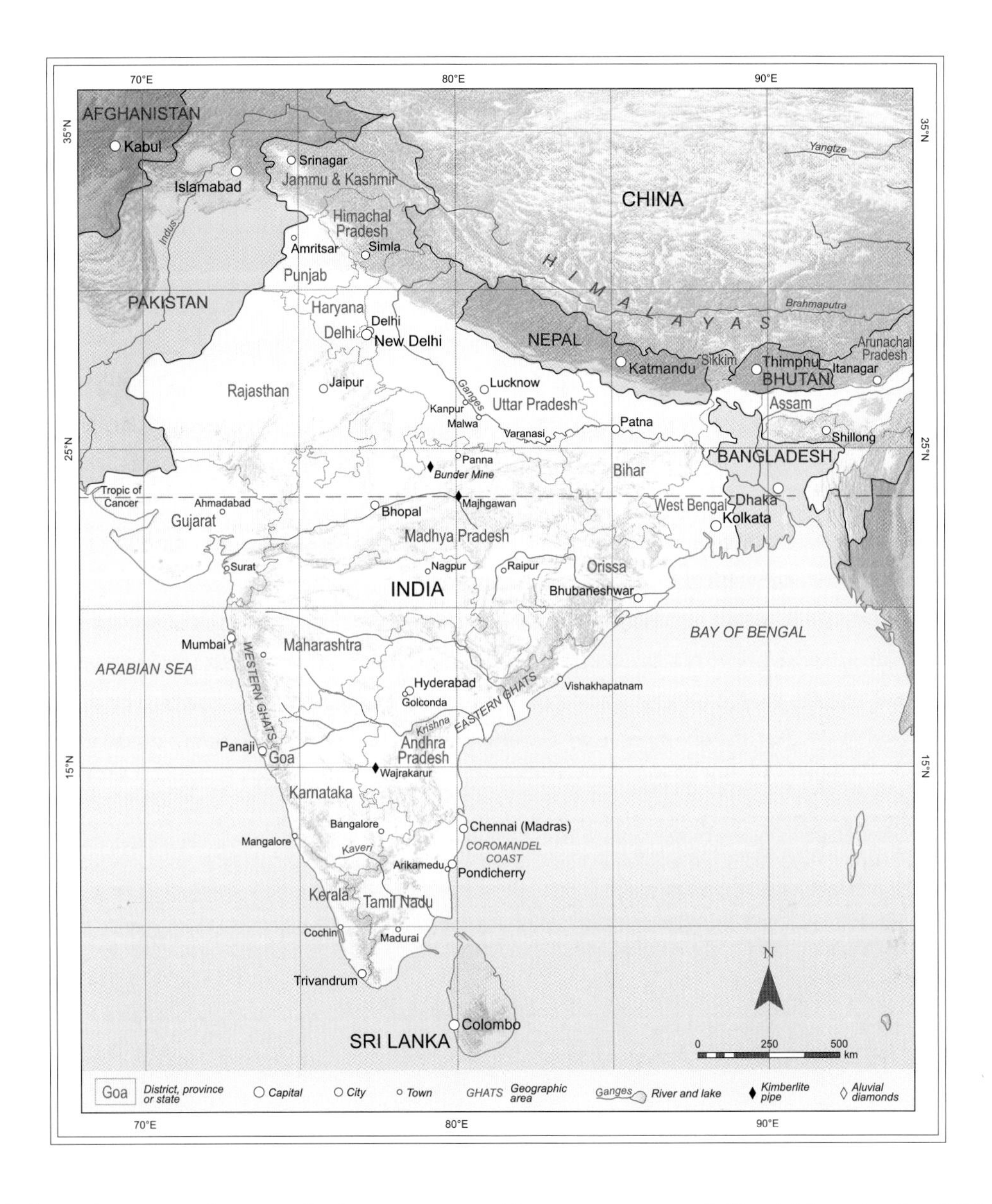

'a force that triumphs over all other powers and becomes master of all neighbouring lands'.[2] It also assured its wearer that he would 'see dangers recede from him whether he be threatened by serpents, fire, poison, sickness, thieves, flood, or evil spirits'.[3] It makes prices of thousands of dollars a carat pale into insignificance.

Buddha Bhatta gave more recognition to the diamond's hardness than had the *Artha-Sastra*, noting its ability to scratch 'even the ruby'. So they remained resolutely uncuttable. In view of this we must assume it was the utter conviction of the Indians that diamonds endowed their owners with a panoply of mystical powers that persuaded the Romans to enrich and protect themselves in a similar way. In its early days the Roman Empire traded energetically with the Indians, for their diamonds and for other goods. An anonymous Roman verse, penned in the second century AD, declared:

> The evil eye shall have no power to harm
> Him that shall wear the diamond as charm.
> No monarch shall attempt to thwart his will,
> And even the gods his wishes shall fulfil.[4]

Pliny the Elder, in the first century AD, could hardly avoid buying enthusiastically into the mystical myth. But, natural scientist that he was, he also noted that diamonds were 'held in great request by engravers, who enclose them in iron, and are enabled thereby, with the greatest facility, to cut the very hardest substances known'. We have Pliny to thank for the first reference in Western writing to industrial diamonds.

An archaeological site at Arikamedu, on the Coromandel coast in southeast India, gives a tantalising glimpse of the first such use of diamonds. Here archaeologists working under Sir Mortimer Wheeler found a bead workshop among other ruins and artefacts. The evidence strongly suggests that the Romans had occupied the settlement at Arikamedu around the beginning of the Christian era, and that diamond splinters had been used to drill holes in onyx, quartz and other extremely hard materials prized for bead-making. The findings are a

We can be thankful we never had to work in the Valley of Diamonds.

salutary reminder to us that European settlement in the East started long before the voyages of discovery by the Portuguese, the Dutch and the British.

And so the narrative strings its way forward sketchily, there being few dates and places in the early centuries to which diamond-related events can be pinned. Legends are legion, though, the most enduring being that of the Valley of Diamonds. No diamond history is complete without it and, it would seem, no record in historical times of Asian exploration could dispense with it. An Arab tract on mineralogy tells how Alexander the Great found his way to a valley in the east of India that was so deep that its bottom could not be made out in the mist. Only by tricking the serpents that guarded the valley with mirrors, so that 'caught by the reflection of their own eyes' they perished, was he able to get close enough to complete his mission. To recover the diamonds known to be at the bottom of the valley, sheep were slaughtered and their flesh hurled down into the depths. The eagles and vultures circling high above the valley wasted no time in availing themselves of the feast thus offered, little knowing – or caring – that diamonds were by now richly embedded in the surface of the flesh. The time had come for Alexander's men to play their part. As the birds settled to enjoy the feast they were dispatched by the soldiers, the gems falling to the ground, the only task remaining to collect them.[5]

The great Venetian traveller, Marco Polo, imparts a note of reality to an otherwise fantastic tale. He landed in southeast India on his return from China in the late thirteenth century, and described a 'kingdom of Mutfili' – his spelling is almost certainly a corruption of Motupalle – in the 'Madras Presidency'. He notes: 'It is in this kingdom [Mutfili] that diamonds are got; and I will tell you how. There are certain lofty mountains in those parts; and when the winter rains fall, which are very heavy, the waters come roaring down the mountains in great torrents. When the rains are over, and the waters from the mountains have ceased to flow, they search the beds of the torrents and find plenty of diamonds.' Marco Polo might have been writing of Lesotho: it is as vivid a description of the first stage of the process of alluvial accumulation of diamonds as one could find, especially considering it was written over 700 years ago, long before the word 'alluvial' was first coined.

He continues: 'In summer also there are plenty to be found in the mountains, but the heat of the sun is so great that it is scarcely possible to go thither, nor is there a drop of water to be found. Moreover in those mountains great serpents are rife to a marvellous degree, besides which other vermin, and this owing to the great heat. The serpents are also the most venomous in existence, insomuch that any one going to that region runs fearful peril; for many have been destroyed by these evil reptiles.'

Marco Polo's description calls to mind the Eastern Ghat ranges that overlook the Bay of Bengal. One of the major rivers cutting the Eastern Ghats is the Krishna, formerly known as the Kistna, at least as celebrated a diamond river as the Vaal and Orange in South Africa. As one follows the Krishna upstream from the coast, inland of the small town of Bezwada, the river enters a tight gorge. It is not difficult to imagine on a summer's day standing alongside the Krishna, looking up into the gorge and seeing, high overhead, vultures soaring. At 16° north of the Equator the heat is oppressive. From time to time there is a rustle in the long grass as animals go about their business, and you wonder where the snakes are. They must be there aplenty.

Whatever its source in historical and geographical fact, the legend of the Valley of Diamonds seems as durable as the stones in its shadow. Much of the early history of diamonds is speculative and fanciful, and it is not until the early modern period that we can be more assured of the accuracy of our narrative. For it was then that developments in Europe took place that would have a direct bearing on India and its diamonds. Starting in Italy, cultural and educational changes seeped inexorably in all directions. Gradually the value of the incomparably hard and bright stones from Golconda shifted. No longer did a tiny talisman hold the power it once had: now diamonds were seen as icons of fashionable adornment for royalty and nobility. From Europe through the Fertile Crescent of the Levant and Mesopotamia over the mountains and into the subcontinent to the south, kings, princes, dukes, mughals, maharajas and sultans bought diamonds to wear.

Ways were found of cutting, cleaving and polishing them which, today, in an age of electricity and lasers, we think of as primitive. We should not be dismissive, though: that they persevered with techniques that were laborious, painstaking and slow is testimony to a passionate determination to enhance the brilliance of their favoured gems. The pioneer craftsmen could be thankful that many of the Golconda gems were large enough for a shaping strategy to present itself relatively clearly and to acquire sophistication quite quickly. In Siberia, Canada or Australia, with their smaller stones, the process might have been far slower.

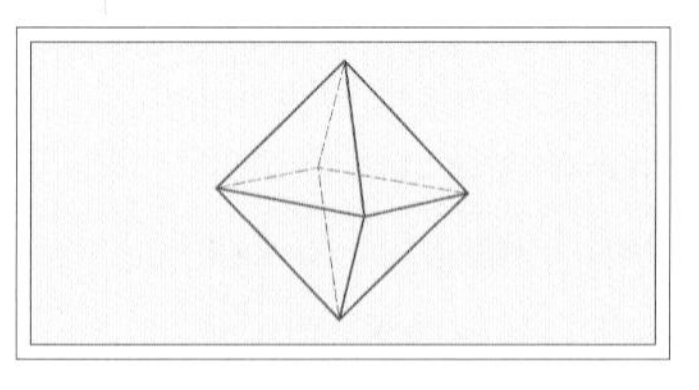

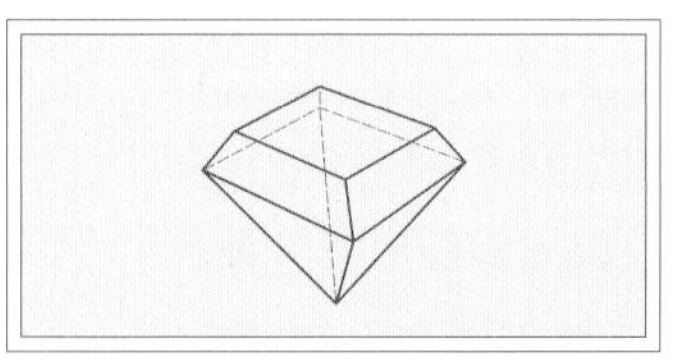

The earliest cuts: the point cut and the table cut.

The first shaping of rough diamonds into gems was the 'point cut', where some cutting was undertaken to perfect the natural octahedral shape of the stone. At some time after the point cut had been perfected, its simplest derivative, the 'table cut', was introduced, which consisted of removing the upper portion of the 'top' pyramid, if an octahedron is envisaged as two pyramids joined at the base.

Towards the end of the fourteenth century, references to diamonds being worked, worn and presented as gifts around Europe gather momentum, and named diamonds start to make their appearance in the records. As the gateway into Europe, both over land and by sea, the Republic of Venice was a natural nexus of the infant diamond trade. Venetian cutters and polishers of those days are recognised as the pioneers of the industry.

The eleven diamonds set in a necklace for Francis I of Burgundy in the early part of the sixteenth century were 'large table- or point-cut diamonds'.[6] Two hundred years earlier the splendid diamonds worn by Henry IV of England on the sleeves of his tunic were point-cut.

Henry IV of England.
(National Portrait Gallery, London)

From the time of Vasco de Gama's arrival in the northwest Indian port of Calicut in 1498, and the Portuguese occupation of what we know as Goa soon afterwards, the slow European penetration of India became inevitable. By the end of the sixteenth century the annual trade in rough diamonds into Europe is estimated at around 18 000 diamonds, sized between ¼ and 10 carats or more. Bearing in mind that only crude cuts were possible long before industrialisation was even dreamt of, and that it was practically only royal houses that were the market for the stones, this is a remarkable figure.

With an eye to spices, silk, diamonds and other resources that the East had to offer, the British formed the East India Company in 1600, and within two years the Dutch followed suit. North and South America had been colonised during the previous century. The great era of global imperialism was well under way.

The narrative of Indian diamonds starts to acquire some credible substance in the second decade of the seventeenth century. At this time the Flemish traveller Jacques de Coutre (who changed his name to Couto in Goa in an attempt to disguise his nationality to the Portuguese, ever suspicious of their Dutch rivals in the Indies) wrote an appraisal of the trading opportunities in India, focusing particularly on diamonds. De Coutre described appalling conditions on the mines he visited during his first visit in 1611, as well as referring in passing to the diamond-bearing terrain: 'The kind of soil in which the diamonds are found are [*sic*] part rocky and part soft which breaks under a little pressure, and are of the same colour as *mengui* with white and black shades. The natives, fifty thousand in all, are very poor, having little to eat, particularly those in the mines …'

A bit later he says: 'Diamonds above seven carats belong to the master and those below to the finder. The workers are always watched for any big diamond.' As to the processing of ground, 'they begin to dig with iron implements and the earth is placed in a sieve made of cane and lined with hide, which they place on the (previously prepared) pavement, till it becomes a heap the size of a man to be spread around for washing. In order to hasten this process they move the earth with their feet the way a farmer does in a field. After this seven to eight men sit on the pavement and beat the heap into powder with square granite stones a span [approximately 23 cm] long.

'When they reach the edge of the pavement, the mud is carried away by the wind and the stones are left behind. Thus the earlier heap of a man's height is now half a yard in height and consists of little stones. The seven then begin to look out for diamonds, big and small, but usually find the latter according to the luck of each. They do not speak while sorting and at times work two or three months without any success. They watch each other while they sort, lest one or the other steals the bigger diamonds. There are guards and merchants at the edge of the pavement besides. Yet they manage to steal and sell it to foreigners for less than half its value. Violations are punished with loss of life and property, which applies likewise to those who buy.'

One aspect of de Coutre's travels that emerges from his writings is the ubiquity of the Portuguese in India at that time: they by no means confined themselves to their west coast colony of Goa. After his visit to Ramanakotha (exact location uncertain, but presumably not far from Golconda), where he recorded the manner of recovering the diamonds already noted, de Coutre visited 'the mine of the Portuguese. Some big diamonds were found here. It was known as such because a certain Portuguese, Alvaro de Mendes, had gone to India in search of diamonds on the orders of King Sebastian. The mine now lay guarded and silent. This caused me to return to Poli where I bought many diamonds and then returned to Goa.'

He was back four years later, visiting mines and buying – and not buying – diamonds. Near his old haunt of Ramanakotha a mine owner, Gopal Raya, 'showed me some big diamonds for sale, of ten, twelve, eighteen, twenty and thirty carats each. But the price he quoted made them unbuyable. Their value in Europe would have been half.'

Two years later, in 1618, de Coutre endured a harrowing interrogation at the hands of the governor of the city of Rachol in Goa, 'the son of an Abyssinian' with 'twelve thousand cavalry with him', when he was wrongly suspected of having illegally bought a diamond of 153 carats. Perhaps it is not difficult to understand the suspicion: by his own admission, 'I had a good many diamonds in the pockets of my trousers.'

Reading of de Coutre's acquisitions in a large number of visits to the Golconda diamond mines, a picture emerges of an area extremely well endowed with large diamonds. The point on which he has been quoted, that diamonds smaller than 7 carats remained in the possession of the owner, gives more than a hint of this. That they worked 'two or three months without any success' would seem due less to the lack of smaller stones than to the primitive methods used, possibly compounded by the indiscriminate choice of which ground to work, though this assumption may be unfair.

Another European traveller from the same century, whose name will forever be remembered with some awe in the narrative of Indian diamonds, is the Frenchman Jean-

Baptiste Tavernier. It was he who brought perhaps the most celebrated diamond of all time – the Hope – from India to the West, though if he is remembered for that and nothing else it would be a pity. He did more than anyone else to popularise Indian diamonds in Europe by selling quantities of them as he returned from his travels there and by his colourful accounts of his time in the diamond fields. Tavernier's introduction to diamonds came when he was apprenticed to a jeweller as a young man. He knew the occasional diamonds he saw came from India and he resolved to go and see for himself where exactly was their source. He began to realise his dream when he made his first visit to India at the age of 25, in 1631, and he was 63 when he returned from his sixth and last expedition there.

Jean-Baptiste Tavernier.

Golconda – close to today's Hyderabad – was the capital city of the Muslim kingdom of the same name, in those times a magnificent and modern metropolis by all accounts, as grand and cosmopolitan as any place Europe had to offer, and a far cry from the humble diamond towns that would later spring up in Brazil and South Africa. Although it was the centre of diamond manufacture and the accompanying jewellery trade, it may have been splendid because it was 150 kilometres from the mess and misery of the mines.

While he had visited India before, Tavernier's first experience of Golconda came in 1642. He was directed from the city to Kollur, close to the Krishna River, where he found the most active mine at the time. He described a mining and processing operation not unlike that seen by de Coutre except that the first step in the concentration of a large volume of mined material was 'wet'. It consisted of emptying the 'ore' into a pit, mixing it with water to form a soupy mixture (called in South Africa 'puddle' or 'porrel'), and draining off the fine mud at the top. The coarser material would sink to the bottom. This would be dried, winnowed, pulverised and winnowed again. In this way all the finer, lighter material would be disposed of, leaving a concentrate of pebbles and coarse sand. This would be scoured through and the diamonds picked out.

Records show that in 1677 all the mines in India produced a total of probably several thousand stones, mostly over 1 carat. From the fact that a minority of stones were good gems there is a suggestion that the mines were close to their mother lode, since as a rule the alluvial process eliminates inferior stones. On the other hand, no pipes have been discovered nearby in centuries of searching, so the question remains open. The high percentage of large stones

A photograph of Golconda today.

raises the question again whether the bulk of smaller diamonds were slipping through the net. It seems likely. From the annual production in the year 1600 of 18 000 stones, it can be seen that Indian diamond mining was on the decline.

Like de Coutre, Tavernier was struck by the masses of people involved in the mining and basic manufacturing – 60 000 in all, including children – and their abysmal living conditions. Some primitive cutting was carried out to eliminate the most conspicuous flaws and inclusions, after which they were valued and sent to Golconda, where they would be faceted further into gems and sold. Though stones larger than 10 carats (as recorded by Tavernier) were strictly speaking the property of the Sultan of Golconda, this could not be rigidly enforced and Tavernier was able to buy a number of large gems, including a rough-cut 112-carat stone that he described as clear and of a beautiful violet. It is generally accepted that this is the stone – violet being the term that would have been used to describe an intense blue colour – which would become the Hope (or Great Blue) Diamond of legend.

There is no record of how Tavernier was able to acquire such a large and unusual diamond, though it was almost certainly from Golconda's prime minister, Mir Jumla. The two became well acquainted during Tavernier's frequent visits to the area and Mir Jumla was generally recognised as being corrupt enough to trade in his Sultan's merchandise on a brazen scale. When he defected from the Golconda sultanate in 1655, to join the eminent Mughal emperor, Shah Jahan, Mir Jumla is said to have helped himself to another legendary Indian diamond from the Krishna River diggings, the 787½-carat Great Mughal, which, after passing into Mughal possession, possibly as part of the famed Peacock Throne, disappears into a mist of uncertainty.

Tavernier reported having seen this stone, as well as the third Indian diamond of extraordinary distinction, the Kohinoor (Mountain of Light). It could be debated which of the Kohinoor or the Hope was the most historic diamond of all time, in the sense of involving themselves in bizarre circumstances and an astonishingly tortuous passage from discovery to ultimate destination. The Kohinoor enters the written record in 1304 when it was owned by the rajahs of Malwa (in Northern India), the Hope 350 years later when it was bought by Tavernier, but both were subjected to a multiplicity of exchanges and are rightly viewed as the consummate embodiments of romantic legend. The hyperbole is due.

To assemble the parcel of diamonds he sold to Louis XIV in 1669 after his last visit to India, Tavernier must have been actively engaged in trading for a long time. Apart from the 112-carat 'beautiful violet', he had stones of 51, 31 and 29 carats, two of 20 carats and eleven between 16 and 5 carats. They were all cut according to patterns not used today; some quite crudely, with few facets, others with a large number of small facets. Tavernier was a regular enough buyer of diamonds to establish a system of valuing stones that is still used, in a refined form, in modern trading. In its simplest terms this uses the principle that diamond size and value are not directly proportional but are related in such a way that stones above, say, a carat have an increasingly exaggerated dollar (or rupee) value per carat the bigger they are.

The 45.52-carat Hope Diamond, which supposedly brought bad luck to its owner, is a popular attraction at the Smithsonian Institution, Washington.

Although in Tavernier's India diamonds had been recovered from the surface or shallow excavations for two thousand years, little was understood of their geology. In the hot, humid tropical climate deep, vegetated soil blanketed the surface: outcrops were rare. It was no country for geological speculation. Wherever diamonds were recovered, though, they were almost invariably accompanied by rounded river stones – pebbles, cobbles and boulders. Sometimes the diggings were on the banks of active rivers, sometimes on terraces a few metres above the water, sometimes high above the river, where the rock was hard, needing to be pounded with hammers to release the fine material in which the diamonds were hidden.

By the time Tavernier left India, diamond production was dwindling. It was not quite done, though. Large stones continued to be found, among them the 410-carat Regent found in 1701 in the Partial mine on the Krishna River. We can be sure that other good diamonds of substantial size were recovered, but clearly Indian mining was scaling down.

So far we have seen the unfolding of the Indian diamond history through a glass, darkly. From about 1800 we start to see it face to face. From 1700, for over a hundred years little was heard or seen of Indian diamonds. After 1730 European diamantaires were able to supply a burgeoning market from Brazilian sources via Lisbon. Then, in the early nineteenth century, as the British dominion in India moved towards colonial rule from a trading presence by the British East India Company, Colonel James Franklin of the 1st Bengal Cavalry found diamonds in the north, in the wide Yamuna-Ganges valley in 1827.

Franklin noted that the diamonds being found at Majhgawan, 19 kilometres south of the town of Panna, came from green calcareous mud, called *muddha* by the miners and identified by them as being distinct from the stony gravel that was the usual source rock. Neither they nor Franklin attached any significance to the difference, however, nor apparently did H.B. Medlicott, who mapped the area for the Geological Survey of India in 1859. Medlicott did record that the diamond-bearing formation extended to great depth, but it was only in 1930, when the name 'kimberlite' was well established in the diamond vocabulary, that the true identity of the Majhgawan pipe was recognised by the Panna State geologist K.P. Sinor.

Conglomerates, too, were mined for many decades in Panna. An account published in 1952 of the mining there vividly illustrates how valuable the little gems were to the local population and the lengths they would go to to recover them, from 'narrow conglomerate beds overlain by up to 70 feet of overburden'. 'To reach the conglomerate, a circular excavation, 150 feet in diameter at the surface and sloping inwards, is made by hand, the debris being carried to the surface in baskets, on the heads of women and children, via a peripheral spiral stairway cut into the side of the excavation. On top of the conglomerate, radiating tunnels 4 feet by 3 feet are driven for a distance of 15 to 20 feet from the sides of the opencast, and interconnected for ventilation. The conglomerate is broken up by moil [a short hand tool with a polygonal point, used for breaking or prying out rock] and by wood or dung fires, and the mined-out areas are supported with sandstone blocks carried from the surface. Lighting is provided by a cotton wick in an open saucer of oil, and water is baled to the surface in small lifts by baskets, or raised by a series of wooden Persian wheels [an *ara-ghatta* is a spoked wheel with buckets on its circumference, powered by draught animals] fitted with calabashes. On the surface the conglomerate is broken fine with hammers, washed by hand to remove the slime, and then dried and hand-sorted for diamonds, without any concentration whatsoever. The average yield is approximately 10 carats per 100 loads (1 load = 1600 lb or 0.8 tonne).

'Each little mine is leased by one or two families, every member of which, from six years of age upwards, spends the mornings in the wheat fields and the afternoons in the mine. Only men and older boys may work in the tunnels, while women, girls and younger boys carry the broken rock to the surface. The mouths of tunnels are draped with Hindu prayer symbols and, lest the yield be decreased, shoes are not worn in the workings, which are regarded as a temple. The red and green shales in the sides of the excavation, and the bright red saris of the women, complete the colourful scene.

'Until the Central Government took office, royalty from the mines was paid to the reigning Maharaja of Panna, who was also entitled to retain the larger diamonds. From among these, the Maharaja's predecessors assembled a necklace of 52 perfect, uncut, octahedral stones, varying in size, up to 25 carats, and aggregating 325 carats, beautifully set in gold.'[7]

Far to the south, in Andra Pradesh, the Wajrakarur pipes – like Majhgawan in the Panna field – were identified comparatively recently as kimberlite, although there is an ancient (almost certainly pre-Kimberley) shaft deep into the No. 1 pipe. It has been tested recently with disappointing results, and the source of the large gems occasionally recovered from the surrounding gravels remains elusive.

The 88.7-carat Shah Diamond was inscribed on three separate occasions, from 1591 to 1824, shortly after which it was given to the Russian Tsar. It is now part of the Kremlin Diamond Fund.

The ancient pipes that have fed diamonds into 800-million-year-old conglomerates in the headwaters of the Krishna River are thought not to be those at Wajrakarur, so perhaps a major discovery awaits patient and persistent exploration. The Banganapalle conglomerates, though nowhere thicker than one and a half metres, have been mined from the earliest days, since the Krishna River alluvial diamonds were traced back to them. Compared, for example, with alluvials mined in South Africa today, their grade, at 2 to 3 carats per hundred tonnes (cpht), was high and the quality and size of stones excellent. Until a pipe producing big diamonds is located in the Krishna headwaters, it must be assumed that stones like the Kohinoor and the Hope and others were washed out of the conglomerates.

S.V. Satyanarayayana is a worthy expert on the Banganapalle conglomerates and Krishna River alluvials. If you are lucky enough to have him as a guide, he will show you the old workings on the conglomerates up on top of the hillside, where you can find lumps of conglomerate that have become rounded by being manually pounded by the ancients to break up and release hoped-for diamonds. If necessary, the local ruler would send in the army and they would seize villagers to join in the labour.

If the Wajrakarur pipes are not rich enough to sustain a commercially viable mine, the same is not true of Panna. Here the National Mineral Development Company mines

ATMAVIRESHWAR STHAPAK

In contrast to most of the distinguished names that grace these pages, Sthapak is at an early stage of what is sure to be an illustrious career. After graduating with a master's degree in applied geology in 1990 from the University of Saugor in Madhya Pradesh, he worked for five years for ACC (Associated Cement Companies), then for the joint venture between ACC and Rio Tinto, before moving across to Rio Tinto. There he was soon a key part of the country-wide search for diamond-bearing pipes of kimberlite and lamproite which the company launched. His outstanding organisational ability, sound geological grounding and exceptionally keen mind saw him rise above the general level. The discovery of the commercially viable Bunder pipe, now in its early stage of development, was just reward for his application. Sthapak is now project manager (diamonds) for Rio Tinto in India, and while Bunder was the first discovery in that country for the miner, given Sthapak's dedication it may not be the last.

a 14-hectare pipe, which runs at a grade of about 10 cpht, with stones valued well above the global average – reputedly US$225 per carat. This is the country's only major producer, though, and, supplemented by a small amount of artisanal output, it gives India an annual production of around 20 000 carats. When you think that Botswana's Jwaneng mine produced 15.6 million carats in 2006, you realise just how small the Panna mine is by international standards.

India's standing in the diamond-mining world will receive a substantial fillip when the new Bunder Mine discovered by Rio Tinto not very far from the Panna Mine comes into production some time between 2013 and 2015. The estimated diamond content of the deposit is 27.4 million carats, and at a grade of around 70 cpht there is no doubt about the project's feasibility. It will be mined at the rate of 2 million carats a year for well over a decade, and as a result India will once again find its way into diamond-mining annals. Hopefully

Rio Tinto geologists at the Bunder discovery site: a few square metres of kimberlite which in the next storm might have been covered by boulders.

during that time more pipes will be found and the country can stand among the big players for a long time to come.

Rio Tinto's presence in India is a reminder of the curiously ironic world that diamonds occupy. In a future chapter we will follow the arrival of the then daughter company Conzinc Rio Tinto Australia (CRA) on the diamond stage, with the discovery of the massive Argyle deposit in the Kimberley Range of Western Australia. We will see how the decision was taken that the bulk of production of this mine should be sent to India for cutting and polishing and how this helped make Argyle an enviably successful mine. First CRA, and subsequently Rio Tinto (when the group's activities were rationalised and the conglomerate restructured), developed a close relationship with the Indian cutting industry.

According to the consulting geologist Chris Smith, 'In 1989 Rio Tinto established its Mumbai office to help enhance the local cutting industry in their processing of the major part of the huge diamond output from the new Argyle Pipe in Australia. The Indo Argyle Diamond Council (IADC) was formed in 1993, whereby Argyle worked together with the Indian diamond industry to promote awareness and acceptance of Indian-made diamond jewellery in the USA.'

At the same time, as part of a global drive to deepen its entrenchment in the diamond business and following its successes in Australia, Zimbabwe and Canada, Rio Tinto applied its

A Rio Tinto customer's modern manufacturing facility in India.

geological and geophysical expertise to India. Its technological pre-eminence was rewarded. Bunder will be a mine, and there may be others in India. And it is very likely that a large part of the Indian cutting and polishing capacity dedicated to the Argyle stones will be turned before too long to beneficiating 'home-grown' stones.

The company is exploring all the Indian cratons, and a large number of kimberlite and kimberlite-related bodies have been discovered. Chris Smith reports as follows on pre-Rio Tinto events on the Bastar Craton in Madhya Pradesh: 'The SE Raipur kimberlites … were discovered in 1988 by tribals (primitive animist aboriginals, some converted to Christianity) living in isolated, forest-covered, hilly, tiger country near the Madhya Pradesh border with Orissa. These remote villagers were supplementing their subsistence farming and hunting by digging for alexandrite gems. They became aware that red garnets were associated with the alexandrite and followed a trail of such garnets in a stream back to a diamondiferous source, thus rediscovering the science of diamond indicator mineral sampling. They illegally worked diamondiferous river gravels and colluvial scree overlying two kimberlite pipes for several years until the unexplained appearance and surreptitious sales of stones in neighbouring towns drew government attention. Investigation by NMDC and GSI [Geological Survey of India] led to the recognition of yellow ground kimberlite in the tribal workings in 1993. Clandestine diggings have continued since, especially at night.'

This informal recovery of diamonds may not carry on much longer. In 2006 the state's Chief Minister said, 'Normally after monsoons, diamond particles flood tribal farmlands. We have reports that rare diamonds are leaving the state through illegal mining by innocent locals.' It was more the plight of the local people and the pitiful price they realised for their transactions that concerned Minister Singh than the loss of rare diamonds.

There are few countries where small-scale illegal mining is not pursued in parallel with multimillion-dollar exploration and mining projects by companies whose names are household words. It happens all over Africa, South America and certainly in India. The scale of operations carried on under the radar is so small that authorities and mining companies turn a blind eye. The listed entities, though, are watched eagerly by one and all in the hope that an important new discovery is in the offing. This is truer in India, with its proud history of diamonds, than anywhere else. Let us hope they will not be disappointed.

Borneo and Kalimantan

Borneo, like India, has produced diamonds for a very long time, a fact as surprising to geologists as it is to anyone who has seen where they have come from.

Remembering Clifford's rule about diamonds being formed under the stable cratonic nuclei of continents and looking at the geology of Indonesia, with its dozens of islands, its volcanoes and its earthquakes, a geologist might say 'impossible!' And he would be wrong. Similarly, if you have been to Borneo or even just seen photographs of it, with its rugged mountains fringed by equatorial forests and mangrove swamps, you would wonder how any sort of surface mining was possible. And we haven't spoken about the climate – the three metres of rain a year falling mostly in a monsoon season that lasts from November to May and temperatures that would leave you wet if the rain hadn't already.

That diamonds were recovered at all from alluvial deposits on the island of Borneo is a tribute to human endurance and determination. But that is only one of the reasons we include Kalimantan in this chapter. Another is that diamonds on any island are a geological curiosity. Moreover, after India's, Borneo's workings are the most ancient. And, lastly, during its heyday in the eighteenth century, when the Dutch mined the deposits, Borneo was an important producer, adding an estimated 50 000 carats annually to the Indian supply.

Borneo may not boast a string of famous gems like India, but in 1965 a 166-carat diamond was found, which came to be known as the Tri Sakti ('three times sacred') diamond. Louis Asscher, nephew of Joseph Asscher, who cut and polished the famous Cullinan diamond, was asked by the Indonesian government to cut the stone. And as if to emphasise how diamonds find their way into blood-lines, the information I have about the gems that were cut from the Tri Sakti comes from John Rudd, great-grandson of Charles Rudd, Cecil Rhodes's partner in Kimberley. Asscher wrote to John Rudd in the latter's capacity as editor-in-chief of the *Industrial Diamonds Quarterly* (or *Indiaqua*), saying, 'The stone was finished in March 1966 and the main one was an emerald cut of 50.72 carats, and further there were 3 marquises weighing 4.04 ct, 2.72 ct and 2.64 ct. They all were of the highest colour and perfect. The big stone first went back to Indonesia, where my brother handed it to the late President Sukarno …'

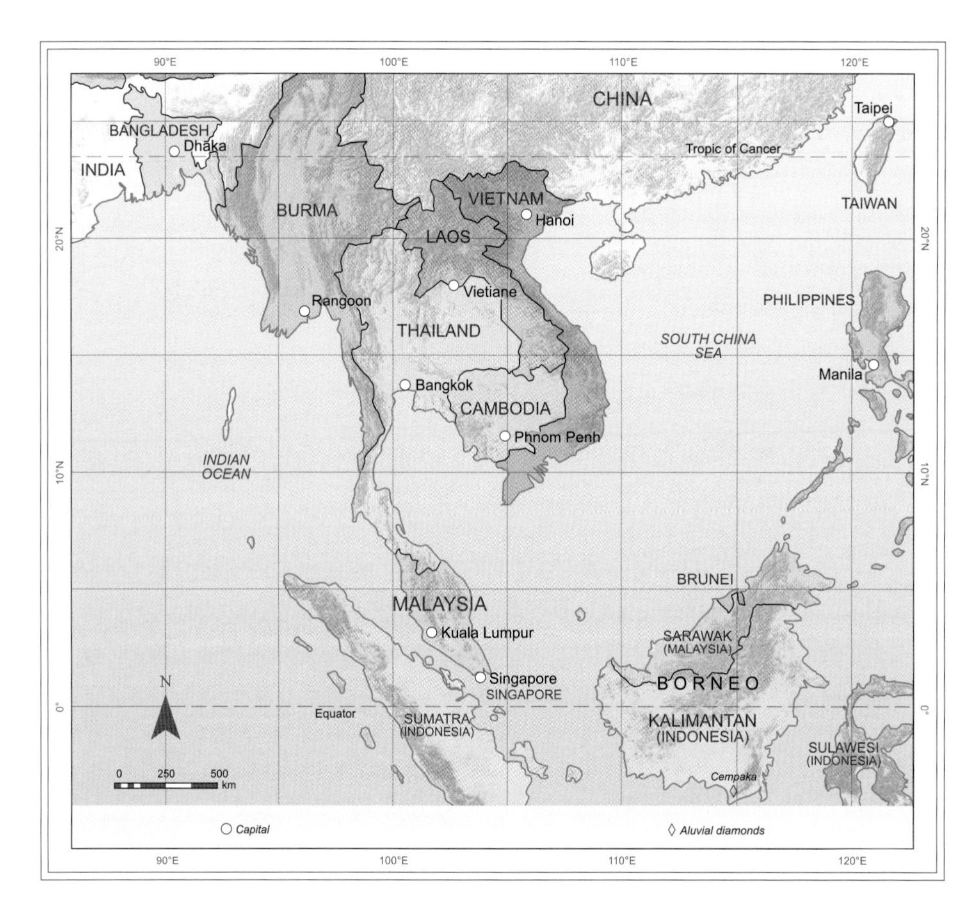

In 1990 a 48-carat stone was found: like India, southern Kalimantan is thus endowed with big diamonds. It is known, too, for its coloured stones, the most common being the fancy yellows, though 'cognac', green, pink and black are also found. In 1987 *Indiaqua* reported on a 1032.34-carat parcel of Kalimantan diamonds from an exploration project with a 6.66-carat stone valued at $2100 per carat and another stone weighing in at 8.53 carats.

These days mining continues on a stop-start basis, mainly bedevilled by the difficulties of terrain and the low grade of the gravels.

To try to understand where these 'island diamonds' have come from, Chris Smith and a number of other world experts from Britain and Australia, including his wife, Galina Bulanova, have studied them, drawing on a formidable pool of expertise and the most sophisticated technology available. They have come part of the way to solving the riddle, being able to say they 'most likely were brought to the surface by lamproites or kimberlites', but where those kimberlites or lamproites intruded is pure speculation. 'The source location … could be at depth within Kalimantan or in other neighbouring parts of Australasia/Asia where deep lithospheric relics may occur' is as far as Chris would go – off the record. Over hundreds of millions of years and through an unknown number of cycles of transport and deposition, the Kalimantan diamonds have hidden the evidence of their ultimate origin beyond retrieval – for the time being at least.

'Pigeon's blood' rubies from Burma.

Burma, Thailand and Sumatra

The problem of provenance is no less puzzling for the diamonds on the Southeast Asian mainland and the island of Sumatra. There one finds a small production as byproduct from mines of other gemstones, mainly rubies, as well as from tin dredging. The same sort of studies as were undertaken by Chris Smith, in this instance by mainly Australian geologists, led to a similar conclusion to that reached for Kalimantan, that 'the diamonds from Myanmar [Burma] and Thailand are derived from normal mantle sources'.

All the diamonds in those places and Sumatra have come from a geological terrain that was joined to northwestern Australia until early in the Permian (around 260 million years ago). Like India, it was part of Gondwana, before, piecemeal, it broke away. There are diamond-bearing kimberlites in the Kimberley region of Western Australia that predate the Gondwana break-up, so this is a possible source area for the Asia mainland and Sumatran diamonds. As with Kalimantan, it remains a vexing question. Diamond geology has a tradition of vexing questions, though, and over the decades a gratifying number have been answered. Perhaps more clues in the Southeast Asian mystery may come to light, and another riddle will have been solved.

I have seen colour photographs of a small parcel of diamonds from a prospect in northern Burma. They are low-value gems, mostly yellow but not the fancy yellow that command high per-carat prices, although the average size of the parcel was a respectable 1 carat plus.

If India is famous for its legendary diamonds, Burma is known for rubies of the very finest colour, a deep blood red with a suggestion of blue. Stones like this, whether or not they were mined in Burma, are described as Burmese rubies or pigeon's blood rubies. They seldom exceed a few carats, but a flawless stone may be sold for millions of dollars.

It seems that Burma would do well to concentrate on gems whose value is enhanced by the intensity of their colour rather than those whose tint counts against them. Diamonds will continue to be won as an occasional byproduct, their origin for the time being obscure. For observers of the passing scene in diamonds, the watching brief we keep on Southeast Asia is more in the hope of an answer than the expectation of seeing a major new mine opening up shortly.

Diamantina today, where it all started nearly 300 years ago.

CHAPTER FIVE

BRAZIL AND VENEZUELA

For nearly 150 years Brazil reigned supreme in the supply of rough diamonds. The discovery of the first stone there in the mid-1720s coincided almost perfectly with the demise of large-scale diamond mining in India; and a century and a half later the Cape Colony walked onto centre stage as Brazil left.

The commonest of several stories of the Brazilian discovery is that a prospector who had spent time on the diamond mines of Golconda recognised the hard white stones showing up in the *bateias*, the wooden pans used by prospectors scouring the gold rivers of Minas Gerais, for what they were. He knew they had more value than mere counters in the card games the diggers played.

As the gold panning continued, more and more diamonds were recovered. A parcel was sent to Rio, and from there by sailing ship to Lisbon. Months after the true identity of the counters was first recognised, celebration broke out in the Portuguese capital. It was common knowledge that the supply of Indian diamonds had slowed to a trickle and the question of where new jewels for the royal regalia of European monarchs and their households would come from had become more vexatious by the day. Jewellers, fretting as they turned customers away, sized up the new stones reaching them from across the Atlantic, in and through Lisbon. Though they were inferior to Indian diamonds, the cutting and polishing houses had no option but to take them, well knowing that no buyer would spurn a brightly blinking stone on the grounds of its provenance.

And then – thankfully – the stream from the East started to regain its old vigour, and diamonds coming from India could again be marketed as such. But there was no new discovery in central India to thank for the restored supply, only Portuguese expediency. Finally the secret that the stones reaching Amsterdam

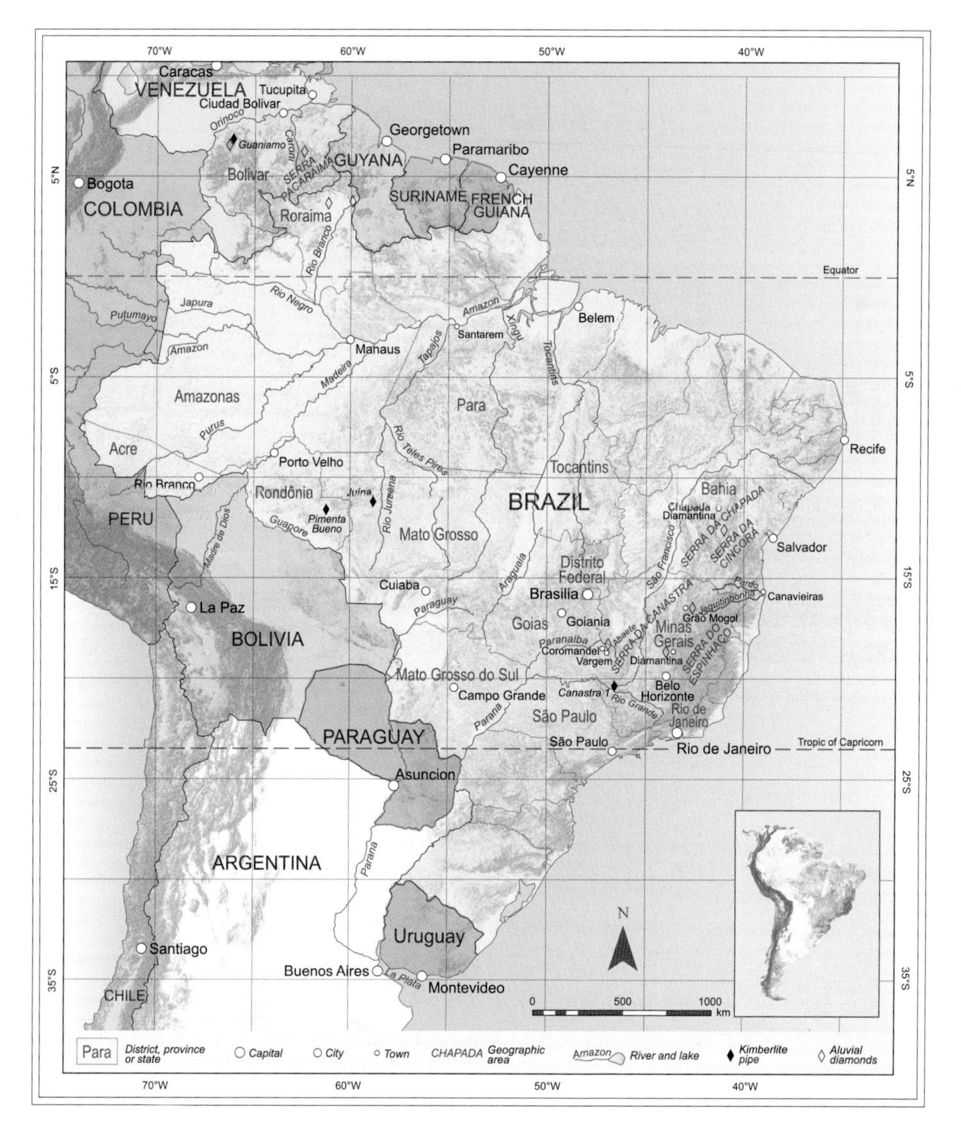

and Antwerp from over eastern horizons came in fact from diggings far further afield could be kept no longer. Allegations of the inferior quality of Brazilian stones were soon forgotten.

Back in the diamond fields of Minas Gerais the outpost of Tejuco, later renamed Diamantina, became the centre of a diamond boom, and the Portuguese Crown took control of the diggings. Rights were ceded to favoured planters, who were quickly able to switch their slave labour force from farming into the often more remunerative riverside occupation of diamond digging.

Lisbon, where the stones were landed, was more than just the seat of government for Portugal and its colonies in South America, Africa and India: it was a diamond centre in its

Lisbon, viewed across the Tagus River, in 1832.

own right. This position had its origins with the establishment of a shipping route to India and the Indies in the late fifteenth century, when Bartolomeu Dias was the first to round the Cape of Storms (soon to be renamed the Cape of Good Hope). Next the Portuguese established a bridgehead into the Indian subcontinent with the founding of a colony at Goa. In consequence Lisbon became the perfect entrepôt for Indian goods, including diamonds, into Europe. In Lisbon there was a community of Sephardic Jews who soon established themselves both in trading and in cutting and polishing the merchandise still trickling in from the Indian fields. Even though the focus moved gradually away to Antwerp, Amsterdam and London, Lisbon retained a hold in the industry, making it well placed to act as a conduit for the Brazilian stones which would soon arrive there in a growing stream. It also placed Portugal ideally to control the industry for a time. And the need for control was soon all too manifest in a trade where price stability had thus far been maintained by the natural balance of supply and demand.

To begin with, Portuguese settlement of Brazil had been concentrated along the coast, with its abundant safe harbours for shipping and readily habitable lowlands. In the steamy northeast sugar cane grew as you watched it and the brazilwood trees so prized as a source of red dye (and of the country's name) seemed to be in endless supply along the coast. Then, in the late seventeenth century a gold boom of unprecedented proportions shifted attention to the southern interior of the country. Thousands of fortune-hunters from Portugal and elsewhere in Europe were drawn deep into the continent, initially through São Paulo. They were intrepid souls, undeterred by the virgin country they encountered, pressing on in search of their El Dorado, leaving little trace behind them. Where rich strikes were made, makeshift settlements sprang up.

Word of the finds reached Europe and imaginations were fired like never before. Young men and old, rich and poor, anyone with a sense of adventure, all booked passages on ships bound for Brazil. Before long the trickle had surged to a flood that numbered many thousands. The first gold boom in modern history was under way. It is estimated that in the eighteenth century around 80 per cent of gold circulating in Europe came from Brazil. Gold was not all the *bandeirantes* (pioneers) found. It may have been what they looked for, and mostly what they found, but it turned out that the hills not only had gold in their soil and rocks; they had diamonds, too.

In 1725 there would have been nothing more than well-worn tracks linking the coast and little settlements like Tejuco; and, along the main rivers, with their fertile floodplain soil and easy water, a sprinkling of intrepid farmers and ranchers. Radiating out from the villages would stretch spider-web strands of prospectors, with their donkeys, basic supplies and the wooden *bateias* that served for gold pans. Tejuco was the hub from which radiated the first winding spokes of diamond exploration, once the stones had been identified on the diggings along the Rio dos Marinhos. This stream is a tributary of the Rio Pinheiro, which, in turn, flows into the Rio Jequitinhonha not far north of Tejuco. It is the Rio Jequitinhonha that has been Brazil's premier diamond river for nearly 300 years, even if others have briefly outshone it during that time.

Panning for gold in the Brazilian mountains, 1823.

Rising 40 kilometres southeast of Diamantina, as Tejuco became known, the river leaves the Serra do Espinhaço, or Espinhaço Mountains, not far to the north of the town, and winds its way northwards across the plains before turning eastwards to enter the Atlantic two-thirds of the way up the coast between Rio de Janeiro and Salvador. Before heading across the rolling lowlands it stays close to the *serra* for some 70 kilometres, with streams that drain the hills joining it at regular intervals. All along this early part of its course there are diamond workings, including Brazil's main dredging operation at the time of writing.

Brazilian garimpeiros *screening for diamonds.*

The first prospectors followed the Jequitinhonha and its various tributaries upwards in search of the source of the stones. Perhaps there was a fabulously rich mother lode dispensing its largesse into the streams. All they found, however, on the slopes was a weathered conglomerate, which, when they panned the softest parts of it, yielded a diamond here and there. Above this layer the streams were barren. Undaunted, they carried on working the river gravels. The yields were not spectacular but the diamonds were of the finest quality and the occasional 10- or 20-carat stone that crowned the glistening heap of dark sieved concentrate at the end of a wash was enough to get them down to the river every day at first light. Every stream running off the conglomerate rewarded their efforts eventually. Once in a while the lucky few working the raised terrace gravels of the ancient river course found traps where the diamonds were concentrated.

From time to time self-styled diamantaires showed up in the centres of mining activity and after bottles of fierce home-brewed *cachaça* had been uncorked and kettles brewed for *cafezinho* – sweet black coffee served in tiny cups – prices would be negotiated, notes counted out and carefully wrapped parcels of stones would change hands. Essential supplies would be stocked up and the digging would go on.

◆◆◆◆◆

IT WAS TOO GOOD to last, though. After five years, as the stones continued to arrive in Lisbon in ever greater amounts, the Crown decided that this was not just a flash in the pan. What they were looking at was an industry in the making and they should be capitalising on it. More than that, after the first consignments of stones had arrived in Europe, prices of 'rough' had stuttered, then started a slow but steady downwards spiral. In 1730 and again in 1739 Lisbon issued decrees that ultimately vested all diamond-bearing ground in the state. Soon after the second decree, ownership of all diamonds sold in Lisbon was ceded to the Treasury. A chief minister appointed exclusive contractor merchants who sold the Brazilian rough wherever they could obtain the highest prices, all on behalf of the Crown. While free enterprise went underground in Brazil, Lisbon became a major diamond-trading and -cutting centre.

Trouble was brewing just over the horizon. In October 1807 Napoleon's Grande Armée invaded Portugal. Lisbon was no longer place for the royal household and court, and under British escort a large Portuguese fleet set sail from the Tagus for the sanctuary of far-away Brazil. For decades the relationship between Brazil and Portugal had been – commercially at least – one of the tail wagging the dog. Portuguese influence in India and the East Indies had been usurped by the more aggressive Dutch and British, and the African 'provinces' acted as a drain on their resources. Meanwhile, in Brazil first gold and then diamonds had more than filled the void left by declining sugar production after the Dutch had toppled the Portuguese from their world dominance. Brazil was still feeling the effects of the boom when the 40-strong Portuguese fleet sailed into Rio de Janeiro's Guanabara Bay early in 1808.

In 1815 Dom João, ruling over the Portuguese empire from the St Christopher Palace in a commanding part of Rio de Janeiro, elevated Brazil's status, constituting it as part of a united kingdom with Portugal and Algarves. Napoleon had been defeated and Portugal had regained its sovereignty. Now there was pressure on the king to come home. Dom João lingered until 1821 before finally bowing to the pressure, leaving his son, Dom Pedro I, in Brazil as regent.

A year later Dom Pedro did what his father might have dreamt of: maintaining the monarchy, he declared Brazil independent. For 67 years, under him and his son, Dom Pedro II, Brazil continued as an independent monarchy. Following a coup in 1889, the country became a federal republic, and has continued as such since then, for a lot of the time under military rule, though more recently with an elected government.

IN THE EARLY DAYS of discovery, for every decree from Lisbon there was in effect an 'equal and opposite reaction' from the diggers. Ironically, perhaps it was this defiant resistance that resulted in the rapidity of the spread of the diamond fields. Each diamond baron well enough established to capitalise on legislative developments, whether by bending or bowing to the statutes, was matched by a hundred diggers – adventurers whose dreams were as big as their purses were small – who slipped quietly across the frontier of authority. The nearest forested ridge unmarked by any cart track was never very far away, and beyond it another swollen river with gravel waiting to be put through their screens.

On they pushed, northwards along the foothills of the Espinhaço range, the Jequitinhonha River never far to the east. Where the river turned eastwards they stayed in the shadow of the hills and kept moving northwards. They then crossed the state boundary into Bahia, where they found more diamonds.

Never far behind the *garimpeiros* (prospectors or diggers) were the concessionaires. Any new find that looked like sustaining a major digging operation was soon claimed as their own exclusive preserve, and it was a matter of days before new contracts were displayed for all to see. In the new find in Bahia their excitement was soon to be doused. Whereas in the state of Minas Gerais (General Mines) digging for minerals was an integral part of the provincial ethos and officials could be persuaded, for a small consideration, to turn a blind eye to illegal mining, Bahia, as the diggers found, was different. This was a proudly agricultural state and fertile floodplains were for growing crops, like sugar-cane or cotton, not for turning over to see what lay below the surface. Bahia's diamond boom would have to wait.

The rugged Serra da Chapada: diamond country.

First the diggers had followed the Jequitinhonha northwards. Now they went westwards, out of its valley into unknown territory. To the west of Diamantina by 170 kilometres one of the world's mighty rivers, the Rio São Francisco, meanders through a wide valley. Four hundred kilometres from its source, it has several thousand ahead of it to where it will enter the Atlantic Ocean near the continent's easternmost point, far to the north. Beyond the São Francisco the diggers found tributaries that repaid the arduous crossing of the great river many times over, while across the next watershed was a hitherto unknown basin, which proved even more richly rewarding. Soon there were diggings up and down the new rivers. Unlike the São Francisco, these, they found, flow southwards, their waters ultimately debouching into the La Plata estuary between Montevideo and Buenos Aires, well south of the Brazilian border.

Soon the diggers' westward push was rewarded, and richly so. They discovered that these gravels yielded substantially more and significantly bigger stones than those of the

Jequitinhonha and its tributaries. In the mid-1700s the biggest diamond thus far found was recovered from Rio Abaete gravels, a 215-carat stone which was named the Regent of Portugal. A hundred years later the Estrela do Sul (Star of the South), a superb 261.88-carat stone, was picked up in the Rio Bagagem, a tributary of the Paranaiba. Cut into an oval gem of 128.8 carats, it was sold to the maharaja of the northwestern Indian state of Baroda, after being exhibited in London in 1862. Four years later, in 1857, a stone of 119.5 carats was found which took its name from the English merchant, E.H. Dresden, who bought it in Rio. The 76.5-carat teardrop-shaped Dresden diamond found its way to Bombay, where it was sold to the same Maharaja Gaekwar.

It was only in the mid-1900s that the main run of very large diamonds emerged from the rivers around the old centre of Coromandel, in what had come to be known as the Triângulo Mineiro, once digging became more or less mechanised and the first dredges were introduced. In quite a small area twelve stones larger than 200 carats were recovered between 1926 and 1949, including Brazil's biggest diamond, the President Vargas, of 726.6 carats, found in the Rio San Antonio in 1938 and sold to Harry Winston. It was followed the next year by the brown 460-carat Darcy Vargas, named after President Getúlio Vargas's wife, who apart from being First Lady was a highly regarded philanthropist. Thus while Diamantina cherishes its status as a World Heritage Site because of its historic importance in the annals of Brazilian diamonds, the accolades for world-class gems must go to the Coromandel area.

Far to the northeast, in Bahia, we saw how in the early days digging for the occasional gemstone had had to stand back for agriculture. The wheel was turning. As the seventeenth century drew to a close, sugar was no longer the white gold it had been, and the brazilwood trees were all but expended. Memories of the rich hauls that had been made far to the south reminded Bahians of their brief foray into the world of diamonds, and now diamonds were being mined just over the Minas Gerais border, in rivers around Grão Mogol. What was more, demand for rough was gaining support from the first ripples being felt of the Industrial Revolution now fully under way in Europe and the United States. The time had come to put Bahia back on the map of diamond producers.

The 726.6-carat President Vargas diamond, about to be cut into many gems.

It was over four decades before the deposits in the Chapada Diamantina or Serra da Cincorá area were found. Discovered in 1844 along tributaries of the Paraguaçu, the

Bahian diamonds dominated Brazilian production for several decades before the main concentrations were worked out.

The Chapada Diamantina mines were extraordinary in one respect: they were the first where 'carbonado' diamonds were found. To this day they remain the main producer of this unusual manifestation of a mineral whose value usually lies in its clarity and 'fire', the ability of a 'white' diamond to refract and magnify light. The carbonado was diamond all right – its hardness left no doubt of that – but instead of discrete crystals, these stones were a mass of tiny black grains stuck together. The agglomerations were slightly porous, so they were not quite as heavy as diamonds of the same size, but, if anything, they were harder. They were thus not without value, particularly in times of scarcity of industrial diamonds from other parts of the world.

Normal industrial diamonds – as loose stones – from Brazil were, by and large, few and far between. Like the Indian stones before them, the Brazilian diamonds came all from gravels, whether in modern rivers, on terraces just above them or in rocks that might be over a billion years old. Wherever the diamonds came from, the source material consisted of a mixture of pebbles, cobbles and occasionally boulders, with sand and clay – and diamonds – filling the pore spaces.

These days, when only a small fraction of global diamond production comes from alluvial gravels, compared with the millions of carats mined annually from kimberlite, it is easy to forget that the old-timers in India and Brazil must have assumed that what they were seeing was a representative spectrum of what diamonds were like. Today we know differently. Of all diamonds that reach the surface of the Earth, it is a minority that are the clear gems which survive the millions of years of rough-and-tumble river transport to wind up in alluvial gravels. Most are small, dirty, flawed and broken. They are the working class of diamonds, destined for the unglamorous world of industry, and they make up by far the bulk of the diamond content of kimberlite mines. To the Brazilians, who had never seen heaps of industrial diamonds, the carbonados would have been a shock.

A much pleasanter surprise awaited the Bahians. In the 1880s, just as the Chapada Diamantina fields were waning in importance, gravels were found along the Rio Pardo, west of the seaside town of Canavieiras, which produced diamonds of the finest quality yet seen in Brazil. So fine they were that it was said that Cape stones from the kimberlite pipes dotted around the new settlement of Kimberley were, in some cases, sent to Canavieiras for forwarding to European diamantaires as coming from their diggings. Whether apocryphal or not, the story suggests that the Rio Pardo stones surpassed the fine Vaal and Orange River stones in their excellence, at least in perception.

From the earliest times diggers from Diamantina, the Triângulo Mineiro and further south up to Mato Grosso in the northwest had sought the mother lode. Even when they

The Canastra 1 kimberlite in southern Minas Gerais gives itself away by its denser vegetation in the grasslands of the Serra da Canastra.

found that the diamonds in the rivers had come from conglomerates up the slopes, they knew that these were just an intermediate stage. It was easy to see that every element, from boulders to fine clay, had been carried to its present position from outcrops of rock in the hills of distant headwaters, in the valleys nearby or on the river banks themselves. If they saw granite boulders in the gravel, they could call to mind outcrops of similar-looking granite further upstream, and the same applied to every other rock-type making up the river stones. It was all straightforward. What wasn't at all obvious was where the diamonds had come from.

The frustration in the *garimpeiro* fraternity had been dramatically amplified when their South African counterparts stumbled upon the source rock for their diamonds almost by accident. What had eluded generations of Brazilian diggers across the length and breadth of the country, the Cape diggers found within a few years, and little more than a stone's throw from the first diggings. Now the *garimpeiros* knew what sort of rock they were looking for.

We are told that the first kimberlite found in Brazil was in 1965, nearly a hundred years after the first photographs of the Cape 'dry diggings' had been seen around the world.[1] Oil geologists in southern Piaui State noticed a pronounced circular depression about a kilometre across and devoid of vegetation. Closer investigation yielded garnets concentrated at the surface but no other diagnostic minerals. Assuming the feature to be a kimberlite pipe, which was named Redondão, they later systematically sampled it and a typical suite of kimberlite indicator minerals was recovered. Diamonds, however, were conspicuous by their absence. Moreover, the find itself was an accident – a lucky and significant one, but nonetheless an accident.

What it did was to establish beyond doubt that there were kimberlites in Brazil. It had been suspected for decades that they must be there, but for all that time not a single pipe had been found. Now the Triângulo Mineiro around Coromandel, a prominent historic source of large diamonds, which would probably not have travelled far from their source, became a priority target area for kimberlite search.

In 1977 the first pipe was found at Vargem, 20 kilometres southeast of Coromandel, but it proved devoid of diamonds. A Cretaceous age was determined, the same sort of age as the Kimberley pipes. The Brazilians gave a sigh of relief. Other pipes would surely be found which had been punched up through the ancient sediments and were not, like the presumed older source pipes, buried by them. And if pipes were found, in an area of widespread diamond distribution, big, deep mines, like those in South Africa and the Soviet Union, were only a matter of time. The pace of exploration was stepped up.

Though more pipes were found, every sampling result sheet was viewed with groans. It wasn't for want of a systematic programme of exploration. Starting in 1961, a joint venture between the French company SOPEMI, with government backing, and De Beers had barred no holds. With both partners boasting a wealth of expertise in diamond exploration in tropical and equatorial terrain from their time in Africa, the quality of the work was beyond reproach. What was more, the exploration was technically a great success in that it found a large number of kimberlite pipes. That was not, however, the news shareholders waited to hear, no matter how hopefully the reports were couched.

By now Clifford's rule was being universally applied, particularly by a company like De Beers, and due attention was given to where the geological map of Brazil showed cratonic nuclei to be present. Not only was the exploration methodology as well honed as could be, but the search was equally scientifically focused: on the cratons.

Outside Brazil the diamond industry was on the threshold of a drastically new era. In 1977 (the year of the discovery of the Vargem pipe by De Beers), across the Atlantic Ocean Falconbridge found kimberlite pipes under the Kalahari in southern Botswana. At about the same time the Superior Oil Falconbridge joint venture embarked on a kimberlite search in the vast cratonic area that takes up most of Canada. This would lead to the single great diamond rush of the twentieth century, in the Canadian Arctic. And exploration was well under way in Western Australia; within a few years it would position CRA (Conzinc Rio Tinto Australia) – and later its parent, Rio Tinto – at the forefront of emergent big players in the industry.

In Brazil it was well known that De Beers and SOPEMI had been exploring there since the early 1960s, dedicating a generous budget and the best possible expertise to finding diamond-bearing kimberlites, without success. Against that background companies asked themselves what hope they had without fifty years of experience in a highly specialised field. In fact, the mining companies were happy to leave the Brazilian field open to De Beers, busying themselves instead with finding gold and base metals and iron ore, all there in abundance. The diggers, spurred by continuing finds of big or coloured diamonds, pursued the alluvials along the tried and tested rivers.

For the academic geologists of government institutions and universities the newly discovered kimberlites in Brazil offered an endless field of study. It had been shown what a window into the upper mantle kimberlites offered, and for the first time Brazilian geologists were given a glimpse of a world deep beneath their feet. Their first prize, though, would be to point future diamond explorers in the direction of a new generation of mega-mines in kimberlites. The Soviet geologists had done it in recent memory; surely they could. All they needed to do was show where the millions of carats of diamonds won from alluvials had had their sources. While De Beers flew instruments in aircraft and analysed thousands of garnets and ilmenites in its laboratories, they would use geology. It was to no avail.

If the kimberlites were, to all intents and purposes, sterile, where had the diamonds come from? everyone asked. With geochemical analysis of kimberlite indicator minerals now within the scope of a number of the bigger laboratories, garnets, chromites and ilmenites from the Coromandel pipes were analysed in their hundreds. All showed that the minerals had formed within the diamond stability field in the upper mantle. Either the diamondiferous parts of the pipes had been eroded away or not all the pipes had been found.

No, said the sceptics, the diamonds had come from far afield, and had been transported to Coromandel by ice sheets. Glacial tillites were well known from different parts of the geological record. To have been spread over such a vast expanse of Brazil, parts of which were not underlaid by potentially fertile craton, the diamonds must have been distributed by ice. What seemed obvious to some was incredible to others. While a friendly war was waged between the academic geologists, exploration teams broke new ground in search of just one fertile pipe.

Summer gave way to winter, carnivals in Rio came and went, inflation climbed and fell. New papers were published in learned journals. Every year new kimberlites were found, some sparsely diamondiferous, most not. In the end those that had lifted the De Beers geologists' hopes for a while were found wanting and farmed out to third parties.

As the 1980s ended and a new decade started, there were new parties looking for kimberlite prospects. Rio Tinto, which had outflanked De Beers in Australia where it had discovered a massive diamond resource, thought perhaps it could repeat the achievement in Brazil. When it took up the kimberlite search in 1990, apart from carrying out its own grassroots exploration, the company entered into a dialogue with De Beers. Perhaps there were properties that the diamond giant had negatively evaluated which might benefit from a completely different approach. One such prospect was an exploration concession at Juína, in Mato Grosso, which De Beers had abandoned in the late 1980s. Rio Tinto felt there was untested potential in the area and, starting in 1972, the British miner applied state-of-the-art geophysical and geochemical techniques to the project.

Glacial tillites commonly consist of a mélange of large, sometimes angular clasts (paler) in a ground mass (mauve) which is much finer-grained.

Rio Tinto had gone in with mixed feelings. While it was impatient to test its theories on how more kimberlites might be found on the claims and – hopefully – turned to account, it was mindful of the socio-political sensitivity of the area where they were located. The claims went to within a few kilometres of the eastern boundary of an indigenous reservation, already well known in media circles, belonging to the Cinta Larga Indians (*cinta larga* means broad belt, the name given to the wide bark sashes worn traditionally round the torso), unquestionably the most widely known indigenous group in Brazil.

Year after year AGMs at the group's plush head office at 6 St James Square in London were disrupted by owners of a single share who represented environmental and political lobbies and NGOs: this sort of thing was grist to their mill. The board gritted its teeth and prepared to defend its stance. If the venture brought South America's first major kimberlite mine into the company prospectus – it seemed like yesterday the same thing had happened in Australia – it was worth the scrap. It was not the closeness of the new Mato Grosso claims to the edge of the reservation that concerned Rio Tinto most; it was the history of the tribal lands. Though the Roosevelt reservation, lying in the western block of the Indian Reserve, where most of the historical trouble had arisen, was far from the exploration targets, care and diplomacy would be called for at all times, and a keen awareness of boundaries.

The Cinta Larga Indians

An aerial view of the large-scale desecration by garimpeiros *of the Roosevelt reservation, homeland of the Cinta Larga.*

Story has it that the Cinta Larga shadowed Theodore Roosevelt's epic trip down the white waters of the Rio da Dúvida (the River of Doubt, which would be renamed the Rio Roosevelt) in 1913 after he had lost the US presidential election the previous year. It would be fifty years before the Cinta Larga appeared in the press, though. In 1963, in what became known internationally as the Massacre of Parallel 11, one of their villages was attacked by rubber tappers, first from the air and then on the ground, with an undisclosed number of deaths. Outrage rippled around the world, but it was decades before the peace-loving Cinta Larga reached the end of their tether. Starting in the early 1700s, they had watched the resources of their homeland plundered – rubber, hardwoods, gold and, lastly, diamonds. It was not so much the exploitation of their natural resources they objected to as the raping of their daughters and the drugs the *bandeirantes* brought with them.

It was after Rio Tinto left Juína in 1997 that the Broad Belt finally snapped. In 1999 diamonds were discovered in the reservation. Indigenous reservations had long since been explicitly excluded from exploration and mining activities, unless conducted by the Indians themselves. Notwithstanding this, after the discovery had been made, the mineralised rivers were overrun by *garimpeiros* who spurned the requests, polite at first, to leave. In 2004 a confrontation between miners and the Cinta Larga reached flash point. Within a short time there were 29 corpses on the ground, though the fatalities were thought to have been a lot more.

GUILHERME GONZAGA

For 140 years Brazil was the world's main producer of diamonds and for decades the diamond giant De Beers searched for their primary source without success. Pipes were found, some with diamonds, but none with the necessary profusion to make a mine. Where the diamonds of history had come from seemed a mystery too tough to crack. Most geologists gave up and moved on to less defiant problems. Not Guillerme Gonzaga. Year after year he has pursued the answer, his determination never wavering. His multiple investigations have suggested to him that the vast ice sheets that ground across the countryside in the geological past stripped the tops off the kimberlite pipes that had been intruded before then. As the ice melted and water became the principal agent of distribution, the diamonds were concentrated. The theory has its detractors, but, undeterred, Guillerme continues to find clues and build his case. He deserves our admiration.

In the Juína rivers, too, there were diamonds, lots of them, but these were not in tribal lands. It was estimated that in the 1970s and 1980s the river gravels of the various rivers in the district produced 5 to 6 million carats per year. This might have done more to put Brazil on the map as a major producer in those times, except for one thing: the diamonds were generally of poor quality. This was not the first case – nor the last – of river diamonds not being of the best. Normally, as along the Jequitinhonha or the rivers around Coromandel or, further afield, the Vaal and Orange rivers in South Africa, alluvial mines can guarantee top prices for their stones, all the dross having been smashed up and winnowed out by high-energy storm-water. Occasionally, though, where alluvial accumulations are close to a very rich source of poor stones, as at Mbuji-Mayi in the Congo, the short distance of transport allows diamonds to collect that would not normally survive the arduous testing of the river environment. Juína is such a case.

This explains, too, why Rio Tinto failed to start Brazil's first big kimberlite mine. Juína exemplifies as clearly as any deposit the uniqueness of diamond mining and exploration. It is not only the size and grade (in carats per hundred tonnes) of the potential ore body that are important; it is, equally, the dollar value of those carats. Rio Tinto could not get the figures to work for a 16.5-hectare pipe at Juína, which is large, with a grade of 40 carats per hundred tonnes (cpht), which is high. Compare that with Letšeng-la-Terai, where the main pipe is smaller than the Juína kimberlite and the grade around 3 cpht: the Lesotho mine is very profitable and Rio Tinto would love still to own it. The average value of Juína diamonds is a few dollars per carat; of the Letšeng stones, depending on the diamond market at the time, around $1500 per carat.

Parcels of Juína diamonds, showing that the rivers will sort out some better stones from the majority, which are small and discoloured.

The company that took the Juína project over from Rio Tinto carried out specialist studies on the kimberlite minerals from new pipes they discovered, with fascinating results. Analyses of a whole suite of minerals – excluding garnets, which is unusual – have shown temperatures and pressures consistent with a depth of 660 kilometres, right at the base of the mantle. An age of 93.6 ± 0.4 million years places their formation at the time of the opening of the South Atlantic Ocean. It has been suggested that 'the origin of kimberlitic magma is [due to a] super-deep subduction process' as the newly formed South America rode over sinking Pacific plates.[2]

On the other side of the reservation that is home to the Cinta Larga, Rio Tinto found kimberlites near the town of Pimenta Bueno in Rondônia. De Beers had found the first pipes during their tenure of ground between 1974 and 1982. Intermittent work had been carried on until 1993, when Rio Tinto acquired the rights. By 1996 its teams had identified 13 new pipes, with varying diamond contents, and stones very substantially better than those just to the east. After rigorous testing, however, none of the pipes was found to have a grade that would repay mining. How often Rio Tinto must have cursed its bad luck. At Juína it was the dollar value per carat that killed the project; here it was the grade. Had these pipes had a 40 cpht grade, they would have made a successful mine anywhere, even in the heart of South America. But it was not to be: the diamonds were too few and far between, unpayably so.

IN A COUNTRY bigger than the continent of Australia with a history of European occupation twice as long, it's easy to see why no clear picture emerges of Brazil's diamond history. The stones have been found up and down its length and across its breadth, yet in its 280-year history it has probably not produced more carats than Botswana's Jwaneng Mine produces in a few good years. I say 'probably' because one can't really say what effect illicit diamond buying has on the statistics. Small stones with large price-tags are the easiest things to smuggle, with, as often as not, the best reason. In Brazil it's a problem that started as the industry was born and Lisbon sought to regulate diamond mining by remote control in a colony with practically no infrastructure. Today it is little better.

Some diamonds from Vaaldiam Mining Inc's Duas Barras Mine on the Jequitinhonha River north of Diamantina.

In March 2006, the NGO Partnership Africa–Canada released a report entitled *Fugitives and Phantoms: The Diamond Exporters of Brazil.* In it frequent reference is made to investigations of the certification of diggings and diamond exports by the Federal Police and collusion in illegal activities by the National Department of Mineral Production. In addition to this a number of perpetrators of fraudulent production are named, including the country's second biggest producer, in Juína. While the objectivity of such a report may be questionable, the wealth of factual data contained – names of people and places, dates, values of production, counterfeit permits and so on – is unlikely to have been invented. It presents an industry with as little structure to it now as there was 280 years ago.

Venezuela

My introduction to diamonds in Venezuela was made through a book entitled *Diamond River,* published in 1963. Its author, an Italian entrepreneur and adventurer, Sadio Garavini di Turno by name, having lost his hard-earned savings in a farming venture in upland Venezuela, hears from a friend of a river in the south of the country 'chock-a-block with diamonds'. With nothing to lose he decides to try his luck and sets off for the legendary river, where he is befriended by a group of Indians.

His account, while undoubtedly rich in hyperbole, is a delight to read. For example, he describes the Caroní River, with 'its surface a vivid blue, with the forest – not to mention orchids of every colour under the sun – reflected in it'. I have travelled up and down a few rivers in western Guyana and di Turno's description rings no bells. Luckily I avoided his encounters with anacondas. Whether or not he did, during the three months he searched, find diamonds 'in almost every dig, and on both sides of the main river', including a 'thirty-six carat beauty', is not the issue: it seems likely that they did find an abundance of diamonds, including some large stones.

In La Gran Sabana highlands in southeastern Bolivar State diamonds have been won for a hundred years by *garimpeiros* and small syndicates from the multiple rivers that drain the Serra Pacaraima along the Brazilian frontier, ultimately to feed into the mighty Orinoco River far to the north. South of the watershed and the international boundary, the state of Roraima in Brazil still sees a small amount of – mostly illegal – diamond mining. The capital, Boa Vista, is used as a 'clearing house' for diamonds from this corner of Brazil, as well as Guyana and Venezuela. The region is extremely remote and the pervasive rainforest and dense network of rivers that bedevil road access make it hard to reach from anywhere, even now.

It was unquestionably much more so a hundred years ago, but with the push for gold and diamonds that had lured Brazilians from Diamantina for two centuries, it was probably opened up from the Brazilian side, with little regard for vaguely perceived international frontiers. Today small-scale digging is carried on intermittently in all three countries across the Roraima highlands. These have given their name not only to South America's highest mountain away from the Andes, the 2810-metre-high Mount Roraima, but to the geological formation from which the diamonds are washed into the modern rivers.

The 1.7-billion-year-old sediments that make up the Roraima Group contain numerous beds of conglomerates, with those at and near the base carrying diamonds. In this respect the area resonates with those in Bahia and Minas Gerais, with their diamond-bearing conglomerates high on the hillsides and on the plateaux. In all of them the kimberlites which brought the diamonds to the Earth's surface continue to elude the most sophisticated probing that technology, industry and academia can dream up.

Indications are that the source of the gold and diamonds in the Roraima Group lay to the northeast of the basin and were washed in by rivers that drained towards the southwest all those aeons ago. It is possible that the source area has been separated from the Roraima basin by the Atlantic Ocean. Let us not forget that this opened up long after those rivers tumbled down from gold- and diamond-bearing highlands, across coastal plains and into an ancient sea. We should look in West Africa, then.

Unfortunately the reconstruction of Gondwana failed to deliver the solution, in that most of the known productive kimberlites in West Africa are far younger than the Roraima sediments. The search is not over. An article in the journal *Economic Geology* offers an intriguing insight. 'Also of interest are the several discoveries of alluvial diamonds in the piedmont region of North Carolina and Georgia ..., whose sources have not been discovered. Could these, as well as the Roraima diamonds, have as an original source the enigmatic and little understood landmass that lay between northwest Africa and northern South America?'[3] Has a vital piece of evidence disappeared: the 'lost Atlantis'?

Southeastern Venezuela is known for the tepuis, *table-top mountains or* mesas, *where the ubiquitous rivers have carved into resistant Roriama sandstone.*

Like the diamonds across most of Brazil, the ultimate source of Guyanese and at least a part of Venezuelan production is an ongoing mystery, defying the direct unravelling afforded diamond explorers elsewhere in the world.

Venezuela did, however, offer one case that was easily solved, at a field known as Guaniamo, in the western lowlands of the country. Not only was this highly productive field, on a tributary of the Rio Guaniamo called the Quebrada Grande (Great Gulley), quite a late discovery, the first diamonds being found in 1970, but it was only a short time before the kimberlite source of the river diamonds was found. In 1982 R. Drew and the South African Baxter Brown, consulting for a group of entrepreneurs from South Carolina partnered with Venezuelan industrialists, discovered deeply weathered and metamorphosed kimberlite at the base of a *garimpo* (digging) being worked. A whole system of kimberlite dykes, sills and 'blows' on dykes was subsequently found in the headwaters of the Quebrada Grande. Having produced an average of around 60 000 carats a year before 1970, after the Guaniamo alluvial diamonds were discovered the annual production shot up to about 350 000 carats for the next five years, though since then it has declined somewhat. Venezuela has never been a major producing country and is unlikely ever to enter that league in the future, but Guaniamo will continue to produce clear gems of not much more than a carat. Any larger stones to have come out of Venezuela, like the 155-carat El Libertador, found in 1942, have been mined from gravels derived from the basal Roraima conglomerates.

At the time of writing, President Hugo Chávez has declined to comply with the requirements of the Kimberley Process, so uncertified diamonds are sold under the table.

Colesberg Kopje, before the 1871 discovery of diamonds on the koppie, which would lead to the opening of the Kimberley Mine.

CHAPTER SIX
SOUTH AFRICA

For aeons San Bushmen have hunted and gathered on the plains of southern Africa. With ancient skills and a sharpness of senses long since lost to modern Europeans, they would have noticed the little stones that shone more brightly than any other as they wound their way along the banks of the rare rivers that crossed their savannah. They may even have discovered their exceptional hardness and used them to shape implements.

The creole community of Griqua living in the present-day area of Hopetown were also sharp of eye and almost certainly were aware of the especially clear, bright stones. So when in 1867 the farmer Schalk van Niekerk let it be known that he would pay a reward for any such stones brought to him, the local people would not have been surprised. Van Niekerk knew they were worth finding. The first diamond he had sold had earned him a tidy packet, £250, and he was sure there were more, many more.

His acquisition of the first stone had been opportunistic. When he spotted it being used in a game played by the children of his neighbour, Erasmus Jacobs, and their friends, Mrs Jacobs had laughingly refused any payment for the stone. How could a child's plaything picked up on the veld have any value? she asked.

Van Niekerk's interest in stones had been spurred by a book he was given by a surveyor whom he had befriended. Among other minerals, it described diamonds. Van Niekerk believed the Jacobses' stone, now his, was this ultimate prize. So did John O'Reilly, a travelling merchant who passed by Van Niekerk's farm De Kalk on his way to Hopetown, some 30 kilometres to the east. O'Reilly was entrusted with the stone and showed it to traders in Hopetown. They pooh-poohed the notion that it could be a diamond, forcing O'Reilly to take it further, to a bigger town further south, Colesberg. The apothecary in Colesberg wasn't sure, but knew who could tell them.

Dr William Guybon Atherstone was a man of many parts. First and foremost a respected medical doctor resident in the regional centre of Grahamstown, he was also an avid collector of fossils and an amateur geologist and mineralogist of some repute. He was to be immortalised in all three fields. As a general practitioner, he was the first outside America and Europe to carry out an operation under anaesthetic in 1847. Two years before, in 1845, Atherstone and a colleague had discovered South Africa's first dinosaur, *Paranthodon africanus*, which they named the 'Cape Iguanadon – bigger than an ox!' But the reason he is celebrated in these pages is that it was he who identified the stone sent him by the Clerk of the Peace in Colesberg as a diamond. He describes its arrival thus: 'Listlessly opening [the letter] in a half-dreamy state, something fell out on the grass. I picked it up – a small, dull, rounded water-worn pebble apparently such as one might find anywhere along the banks of the Vaal or Orange rivers. No outward mark of crystallization or other indication of value, being loose in the envelope which was neither sealed nor registered, a thin film of gum alone preventing it from falling out of the postbag and being lost forever.'

Atherstone set about examining the stone. He first tested it with a file. It would not mark, so it was hard. By weighing it in air and water he found its specific gravity, a third as heavy again as quartz, which might otherwise quite easily be mistaken for diamond. So quartz it was not. Topaz, though, has the same specific gravity as diamond and may be colourless: could it be that? He knew that topaz was softer than ruby, so if the stone in front of him would scratch his ring, it could not be topaz. He dragged the loose stone across a corner of the one in his ring: it left a clear scratch mark. 'Hurrah: it must be a veritable diamond,' he noted, his excitement manifest, the marked ring forgotten. He weighed it again: 20½ carats.

The diamond was sent to Richard Southey, Colonial Secretary in Cape Town, 'in a tin box under seal … in the Governor's Despatch bag'. In his report back to the Clerk of the Peace in Colesberg, Atherstone noted that the stone 'should be worth about five hundred pounds, and that where that came from there must be thousands more'. He was – surprisingly – spot on with the value, as this is what O'Reilly and Van Niekerk received for the stone. But the more remarkable thing in the note was how prophetic his prediction of 'thousands more' was. Had he said millions he would have been ridiculed – but correct.

The first South African diamond to be widely celebrated was to take its first steps on the road to fame falteringly. When it arrived in London, the well-known Haymarket jewellers, Garrards, valued the stone at £500, then washed their hands of it. The *London Illustrated News* would not publish a picture of it and at the Exposition Universelle in Paris, where it had been hoped that the stone would hold pride of place in the Cape Colony stand, a replica was exhibited instead. Sir Roderick Murchison of the Museum of Natural History in London authoritatively dismissed the notion of diamonds being found in the Colony as poppycock. Van Niekerk, though, knew where the stone had come from and, having received his half-share of the £500 from O'Reilly, resolved to find more.

In March 1869 he was rewarded. A local shepherd showed him a beautiful clear white stone the size of the top joint of his finger, which he knew was a diamond. He had to have it. His disposable assets at the time included 500 sheep, 10 head of cattle, a horse and its saddlery (the total value of which was estimated at £150), and probably very little else. For a stone he could comfortably hide in the palm of his hand, he parted with all of his earthly goods without hesitation. Days later he had sold the stone, now weighed in at 83½ carats, to merchants in Hopetown for £11 200, enough to buy him a big farm and all the stock and saddlery he could want.

The Star of South Africa, set in a pendant with smaller diamonds.

Unluckily for them, the Hopetown traders soon found themselves in the middle of unprecedented court proceedings, as the legality of the original ownership of the stone was contested. It took some months, and a judgment in their favour, before they were able to cut the deal of a lifetime, selling the stone for £25 000. With numbers like that racing along the bush telegraph, it was no wonder that the face of the deep interior of South Africa was about to change drastically and irreversibly.

TODAY IF YOU were travelling the N12 highway from Kimberley to Cape Town, you would bypass Hopetown. This is the Karoo proper. You would just have crossed the Orange River, meandering through its broad, barely perceptible valley. It is desolate country that blazes from November to February and can freeze on a winter night. The plains are scrub-covered, with rare tufts of spiky grass and thorn trees mostly confined to the ridges and koppies. Here and there you may see a flock of sheep or a herd of springbuck. You might be forgiven for wondering if anyone lived in these parts.

Off the highway with its speeding traffic, things have changed little since 1869. At that time this was frontier country, with several territorial borders converging near here. Hopetown, south of the Orange River, was in the Cape Colony, with the Boer republic of the Orange Free State lying immediately north of the river. Further north, the Vaal River marked the boundary between the Free State and the Zuid-Afrikaansche Republiek (South

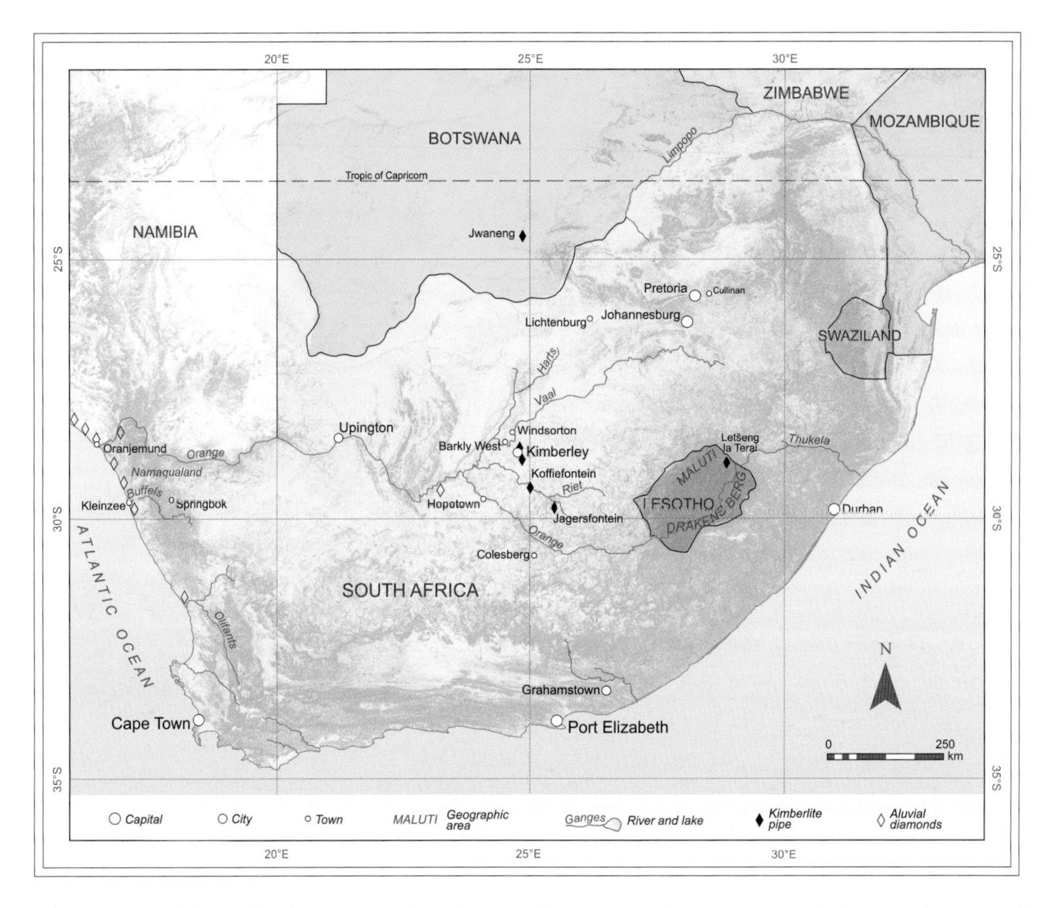

African Republic), also known as the Transvaal. But a wide area around the confluence of the Orange and Vaal rivers had been settled by neither republicans nor colonials but by a pastoral creole frontier community who were to be dignified by the name Griquas, settled around the mission station of Griquatown.

Before the discovery of diamonds, the Cape Colony was down on its luck. According to one contemporary account, 'business [was] generally … in a moribund condition, the aftermath of the Kaffir wars; deplorable stock diseases and severe droughts … being devastatingly felt, and the outlook … gloomy indeed.'[1] Hopetown, founded by local Boers as a Dutch Reformed congregation in the early 1850s, might have seemed optimistically named. Events of 1867 may have given hope to a few, but it was the Star of South Africa, two years later, which would not only prove the name prophetic, but bring relief to the country as a whole.

In 1867 Queen Victoria had been on the throne for thirty years. The Victorian Age, which would reach its half-way mark in the year the Star of South Africa was found, had ushered in unprecedented technological progress and, for some, undreamt of prosperity. The Great Exhibition of the Wares of Industry of All Nations, which the Queen opened in Hyde Park's spanking new Crystal Palace in 1851, was the first of many like it in Paris, London, Vienna, Philadelphia and Chicago. They were as vivid a symbol as any of the direct consequences of industrialisation. The world would never be the same. Steam, both on land and across the oceans, was already shrinking the globe and telephones would hasten the process dramatically before the century was up. For South Africa, and for diamonds, the timing could hardly have been better.

1867: diamonds appear in literature

As it happened, 1867 was a singular year for diamonds for reasons beyond the discovery of the Eureka. It was the year when Wilkie Collins started writing *The Moonstone*, the first classic detective story to be published in English. It is a fictional account of the intrigue surrounding a yellow diamond the size of a plover's egg that journeyed from India to Europe and back again in the early 1800s. By the latter part of the century diamonds had become so deeply embedded in the minds of men and women that Collins was bold enough to hope that his subject would capture the imagination of readers of his new genre.

While Indian diamonds to this day surpass all others in the mystique surrounding them, production from Golconda had all but dried up by the middle of the nineteenth century. That the embryo of an industry had not become a foetus was thanks to a steady supply of stones since the end of the previous century from Brazil.

As the novelist Collins started on his *magnum opus* with an eye to the East, an author of another kind was preparing to document hard facts and considered speculation about the New World stones. The explorer Richard Burton (and the first translator into English of the *Karma Sutra*) visited Diamantina in the state of Minas Gerais in Brazil in 1867, coming away convinced that the country stood at the threshold of a boom in diamonds. As industrialisation, and with it general wealth, gathered pace in Europe and America, so the demand for both industrial stones and gems was burgeoning. Burton was probably justified in thinking the Brazilian fields 'had barely been scratched', and that better organisation and mechanisation would lead to drastically scaled-up production. The young captains of industry in Birmingham and Chicago and their society wives had to be kept glittering, after all. It was not to be. The ancient Gondwana triangle of cratons was about to be completed: from India to Brazil, and now on to South Africa, thanks to men like Van Niekerk and O'Reilly.

In the late 1860s the supply of stones reaching the polishing houses in Europe and New York from the Cape was not much more than a trickle. At first the new stones from the Cape were greeted with sceptical disdain. This was nothing new: as we have seen, the first diamonds from Brazil had met the same fate. Today we know that the diamonds from the Vaal River close to its junction with the Orange and from the Middle Orange, immediately below that confluence, achieve consistently higher average prices than stones from anywhere else in the world. The superlative quality of the first Cape diamonds could not be denied for long.

It was the discovery in 1869 of the 83½-carat Star of South Africa that proved a landmark event. Not only does it mark the beginning of the history of trade and industry in South Africa; it just as emphatically opens the annals of diamonds as a global industry. Within weeks of news reaching the outside world of its sale, men were jumping ship in Cape Town and Durban to join the rush to the interior. Farmers were leaving sons and labourers to herd stock and till the lands; departing lawyers and accountants hurriedly trained clerks to keep their practices going. The timber industry boomed like never before and wagon-makers hired all the hands they could to quadruple their production. Traders in every centre from the ports to the diamond fields filled their warehouses, again and again. The idea of stones worth a lifetime's earnings lying on the surface waiting to be picked up was more than most men – and a few women – could resist.

Crossing the Vaal River at Klipdrift (now Barkly West), the bridge almost complete.

As the focus of the diamond rush, the Orange River around Hopetown soon lost its pre-eminence. Reports of stones found along the Vaal, a couple of days' ride further north, drew the fortune-seekers to spread out from Hebron in the northeast to the Berlin Mission Station at Pniel, 50 kilometres down river. To begin with, there were diamonds to be harvested at scores of places along both riverbanks without so much as turning a spade.

In the country east of the confluence of the two big rivers, closer to the Vaal than the Orange, diamonds had been found where there was no river. Early Boer trekkers had settled close to the sources of perennial spring water, near to which they built their rudimentary clay-brick houses and their stone kraals, their lives centred on these sources of life-giving water. On two of these farms, Dorstfontein (Thirst Fountain) and Bultfontein (Hill Fountain), the owners had found diamonds in yellow clay dug at the surface. The name Dorstfontein was new and never worked its way into local usage, the neighbouring farmers preferring to call the farm by its original well-established name, Dutoitspan. It was this name that would stand the test of time.

Kimberlite rock weathers to soft yellow clay, which would have been as good as any the early settlers could find for their bricks. Thus the story of diamonds found in the bricks of the Bultfontein farm house may not be apocryphal.[2] We are told of walls 'unable to withstand for long the probings to which they were subjected by neighbours, who, with the aid of pocket knives and other handy implements, prised out all the precious gems in sight'.

With the science of geology in its infancy and the origins of diamonds shrouded in mystery, the fact that the stones should have been found away from the rivers drew no particular interest. They were where you found them: what did it matter how they had got there?

A hundred and forty years later, the names of farms given by pioneer settlers – Bultfontein and Dutoitspan – are there for all to see in Stockdale Street, Kimberley.

Today it seems ironic that the stones at Bultfontein and Dutoitspan, which would become two of the family of De Beers mines in Kimberley, were found before the rush to the Vaal River started in earnest. While they would later become the scene of frantic claim-staking and claim-jumping, worse than anything along the rivers, in the early days they were bedevilled by one major drawback. There was – for most of the year – none of the water that made the washing and recovery of diamonds along the rivers quick and straightforward. Not only was the digging easier, but the living too, with shady tree-lined banks and a boundless supply of fresh water for a well-earned bathe. Far from the Vaal, the springs around which farm life revolved could sustain a family, perhaps several, and some livestock, but not a sprawl of diggers, particularly not in the bone-dry winter.

Practical considerations, then, meant that the momentous day when the mother lode was discovered and identified – a new rock as important in its way as any – would not be hastened. For most of the wave after wave of prospectors, the so-called 'dry diggings' were noted and left at their backs with the fabulous Vaal River ahead. The irony becomes apparent when one compares the yield of diamonds from kimberlites and alluvial sources: the former dominates historical global production by orders of magnitude.

Nowhere else has diamond mining evolved as it has in South Africa. The diamond fields in 1869 call to mind the words in Genesis about the world before Creation: 'without form, and void'. Prospectors in those first days spread out and scratched the surface wherever there were reports of stones having been picked up. Soon there were too many diggers in a limited area for unrestricted working, and 30-foot square claims had to be pegged. Then gradually method and system replaced the haphazard digging that had characterised the early days.

On many of the claims, after a metre of surface gravels had been worked through, a hard, dense layer of white limestone – today geologists call it calcrete – was struck, which generally resisted the most determined attempts to break through it with a pick or crowbar. Most often the frustrated digger would move on to another claim. The lucky ones found places where they could break through the calcrete into the gravel below. This might have been loosely bound by limestone or free of it. It could be worked. Mostly it was barely worth

River diggings near the Berlin Mission Station at Pniel.

the effort, and the diamonds were few and far between until the base was reached. Just above hard bedrock, which was indisputably the floor on which the first gravel had been deposited by the river all those aeons ago, the digging was much more rewarding, especially, as they found, if there were big boulders lying on the bedrock and it was potholed and channelled. In some places this basal gravel was extremely rich.

It was the so-called Natal Party of diggers that first saw the advantage of organising themselves. After digging a 10-metre shaft to get to the basal gravel at Hebron for a single 2-carat stone, they moved down river to Pniel where they were more successful, starting the rush to what would become one of the biggest centres of the river diggings. Soon shafts down to the bedrock, and tunnels along it, were common. Only later, once some rudimentary mechanisation was introduced, did it become payable to strip the low-grade shallow gravel off and stockpile it to get to the deeper gravel. Underground mining would give way to open-cast.

By the end of 1870 there were an estimated 10 000 people stretched along 150 kilometres of the Vaal and its tributaries. They were concentrated in the centres of Pniel and, across the river, Klipdrift (later to be renamed Barkly and, soon afterwards, Barkly West) as well as up river at Hebron, which would become Windsorton. Today both Barkly West and Windsorton remain important centres for diamond miners and prospectors.

While life for the diggers was focused entirely on their claims and the tented camps to which they dragged their tired bodies at the end of the day, behind the scenes trouble was brewing. Ownership of the territory which had so quickly become the hottest property on the globe was fiercely disputed between indignant local communities and the Cape colonial government. Seeing an opportunity, the two Boer republics, which had been watching from the sidelines, entered the fray. If 30-foot square patches of gravel were as far as the diggers' claims

Early days on the 'dry diggings'.

went, at a higher level much bigger claims were being staked – for the whole diamond field. In the end, the might of the Whitehall-backed government in Cape Town won the day, the Boers were pushed back, and the Griquas and their cousins found themselves colonial subjects in Griqualand West, originally its own colony but soon to be incorporated into the Cape.

With ever-increasing volumes of low-grade material to be moved to reach the pay-dirt, the diggers longed for ground where the diamonds were not so deep. Some remembered what they had seen on Bultfontein and Dutoitspan, a day's ride away to the southeast. Others recalled stories of diamonds found there in the soft yellow clay just below the surface, with not a boulder in sight. And with good rains having fallen in the summer of 1869/70, there was, for a time at least, enough water if not for washing the 'dirt' – which could be dry-screened if necessary – at least for domestic use.

The diamonds may not have been as big as on the river diggings and the stones included many of inferior quality, but the advantages outweighed those drawbacks. There was little if any hard material to move; it was much easier digging; the diamonds were more consistently distributed as one dug deeper; and there were more of them. The last difference between the river and dry diggings – the diamond 'grade', as it is called by the miners – might have more than made up for the lack of the infinite supply of water nearby. In fact it was more important than they realised. The early diggers would soon learn that the higher grade of the dry ground was essential to keep them in business. Though they were not to know it at the outset, below the yellow clay that yielded so willingly to pick and shovel was hard 'blue' ground, which in some cases would have to be crushed to release its precious gems. And crushing was expensive. In the beginning some of the harder ground was found to decompose quite quickly if left on the surface, but they might have to wait for years to process the denser ground.

Today diamond miners know that a grade of less than half a carat per hundred tonnes (cpht)[3] will sustain a well-run alluvial operation, where no crushing is involved and where most of that hundred tonnes consists of boulders and cobbles which can be screened off before processing begins. In a kimberlite mine, by contrast, where there is no low-cost pre-concentration, and everything mined goes through a complex crushing circuit, the grade needs to be twenty times that of an alluvial mine. In other words, a grade of anything less than 10 cpht is unlikely to pay the bills at the end of the month. The kimberlite at both Dutoitspan and Bultfontein, the first dry diggings to be worked, would be profitably mined by De Beers Consolidated Mines for decades; so it is no surprise that they were a juicier target for hundreds of diggers than unpredictable, often low-grade river diggings.

Some of the first to work the dry ground gave up when they encountered the hard blue ground. Was this not the bedrock that lay at the base of the workable material in the river diggings? One of the fundamental tenets of riverside mining was that the bedrock – provided it was true bedrock and not just a calcrete layer within the gravel – never, ever carried

Life off the diggings becomes sophisticated.

diamonds. Why should it be any different in the dry diggings? Others, more observant, noted that there was not a sharp 'contact' – as geologists call it – between the yellow and blue ground: it was gradational. In places there was a gradual change from yellow through khaki to green then blue. The same glistening grains of mica they'd noted in the yellow material were still there in the blue; or tiny, shiny black grains – ilmenite, an iron titanium mineral that is weakly magnetic – just as prominent in both types. Also, the diggers asked themselves, where were the hard rounded boulders, cobbles and pebbles, of a variety of types, that they had found at the river? No, the dry diggings were different.

While they debated the origins of the diamonds in the evenings, during the day the digging was more frantic than ever. And if the profitability of the Dutoitspan diggings soon overshadowed that on adjoining Bultfontein, there were surprises to come. On a nearby farm called Vooruitzigt (Prospect), owned by the dour, reclusive Johannes Nicolaas de Beer, a new deposit was found in April 1871. It came to be known as De Beers or, after a second, more important locality was found about two kilometres away on the same farm, Old De Beers. The later discovery was on a low rise

The farmer Johannes Nicolaas de Beer never dreamed his name would be immortalised.

called Colesberg Kopje by its discoverers, the Colesberg Party, though the workings soon became known as New Rush.

Considering the months it took to get from just about anywhere to the deep interior of this wide, completely undeveloped country in the 1860s, it stretches the definition of 'rush' to apply the word to the original convergence on the Orange and Vaal rivers after the Star of South Africa had been found. New Rush, by contrast, was well named. There were already thousands of people within a three days' ride by cart or on horseback and the congregation of those numbers in a small area within a few days constitutes a rush by anyone's reckoning.

Diggers at Dutoitspan and Bultfontein had set up their village close to the Dutoitspan Mine, and the New Rush Mine, while close to the Old De Beers diggings, had quickly seen a sizeable settlement spring up around it. In those days one lived in Dutoitspan or New Rush. The outgoing Colonial Secretary, Lord Kimberley, found himself in a dilemma. Vooruitzigt was more than he could wind his tongue around, and New Rush did not sound appropriate in the elegant chambers of Whitehall. What to call the place? The answer came from the Governor in Cape Town, Sir Henry Barkly: he'd like to call it Kimberley. So Kimberley it became, even if the new name took a while to fix itself in diggers' minds, most of whom still called it New Rush long after the change had been formally instituted.

DURING THE 1870s an important realisation began to dawn on the diggers. In Brazil, and in India before it, and much more recently along the Vaal River, all diamond deposits ever worked had been tabular, with never much more than 10- or 15-metres depth to them before the sterile bedrock was reached. They may have made up for their limited vertical dimension by extending over large areas, but there was no question of their reaching deep into the Earth.

The dry diggings at Kimberley were a completely new experience. Not only were they away from the river, but they appeared to have infinite depth. Some of the diggers had come from other parts of the world – California, Ballarat and Bendigo, Diamantina, or even Barberton, just around the corner – and had paid close attention to the rock types where they worked. Many of them had picked up the rudiments of geology. Gravels consisting of smooth, round river stones were like old friends to them, but what was this strange, hard, blue-grey formation they were finding below the yellow clay that carried the diamonds nearer the surface? It contained diamonds all right, as well as a black mineral not unlike granadilla pips in appearance, and here and there clear red rubies, and in places lots of a flaky mineral that must be mica. On their own none of these was strange to them: what was new was to find them all together in the same rock.

That the name Kimberley was hardly dry on the maps mattered not. This rock that had been seen nowhere else on Earth could only have one name: kimberlite. From its mineral

Deeper still: the Kimberley Mine four years after its discovery.

composition and the funnel shape of the bodies it made up, it had to be igneous in origin, come up from the depths. There was no fear of its suddenly dying out: it would go down and down and down. With this completely different geology, what dawned was the startling recognition that the miners were scratching at a formation that would make mines, not shallow one- or ten-man scrapings before they moved on to the next claim, but deep picture-book mines. A vision started to take shape of what Kimberley might become, a picture that only months before would have been beyond the wildest dreams of the most romantic of them.

The De Beers head office in 1898.

The first tacit recognition of the future status of the various deposits was the abolition of the restriction on ownership of more than ten claims. Mines, after all, are owned by big companies, not co-operatives of individuals. Who would own the companies, though? The citizens of Kimberley settled in for a long game of Monopoly.

As the English travel writer Frederick Boyle had prophetically written in 1872: 'These things [diamonds] require the most delicate manipulation, they exact the strictest reticence, they need a hand to hold them back or to loose them as the occasion asks ... By royal monopoly alone, or by means of great and powerful companies, can jewel digging be made a thriving industry. Into the hands of a company all these public fields must fall, and, thus used, they may benefit the country for generations to come.'[4] He might have been writing a mission statement for Cecil Rhodes to use in a few years' time or for De Beers Consolidated Mines in a hundred.

Enter Rhodes and Barnato. Whatever else they might have been, these two Englishmen, so drastically different in background and personality, were among the first mega-entrepreneurs South Africa and the diamond world had seen. Both had been lured to the diamond fields by first-hand accounts of the fortunes to be made there; Cecil Rhodes by his brother, Barney Barnato by a cousin who had preceded him. Neither was among the first claim-owners on any of the deposits, that race having been won by the pioneers of the river diggings. Both were forcibly struck on arrival at the dry diggings by the wealth in circulation, and until the opportunity arose of tapping into the mainstream of this largesse, they were content to get as close to it as they could. That a high profile in the public life of Kimberley would be advantageous to their objectives was as clear to each of them as it was within their compass. It was only a matter of time before they manoeuvred themselves into key positions of both civic power and claim ownership, each recognising the other at the outset as the main competitor in the contest for dominance. To be in Kimberley in the 1870s, watching Rhodes and Barnato jostle for control, would have been an intriguing experience.

Rhodes was nothing if not a consummate strategist. He was content to let the Barnato brothers build up an unassailable dominance in the rich Kimberley Mine, first by securing themselves a block of claims in the prime part of the mine and then by adding to it, claim by claim. Rhodes was astute enough to recognise that there were other early comers in the Kimberley mine with substantial interests there that would present a major obstacle to Barnato if his quest was – as Rhodes knew it must be – to gain control of the mine. This did not apply to the less contested De Beers Mine. Aided and abetted by the genius of his partner in the De Beers Mining Company, Alfred Beit, in May 1887 Rhodes secured absolute control of the De Beers Mine. Barnato, meanwhile, had not been idle in the Kimberley Mine, by now having consolidated three of the four main claim-holding companies under his own control.

To block Barnato, Rhodes knew that his only chance was to buy out the only company in the Kimberley Mine not in the Barnato stable, the powerful French Company. A key card in Rhodes's hand was that this company was represented by Julius Wernher and none other than his own partner in De Beers, Alfred Beit. But Barnato would not be outmanoeuvred so easily. Realising the danger in what Rhodes was up to, he himself acquired a large block of shares in the French Company.

Who owned which claims in the Kimberley Mine, 1882.

Finding himself thwarted in his attempt to buy the French Company, Rhodes appeared to concede defeat. It was a ruse. In what seemed very like a wooden spoon, he negotiated the purchase of 20 per cent of the new reconstituted Central Company that now had absolute control of the Kimberley Mine. Nobody had seen him slip the card up his sleeve as the battle of attrition got under way: the Kimberley Mine versus De Beers; Barnato versus Rhodes. Production was stepped up and diamond prices fell.

Rhodes recognised at the outset that he had to buy control of Barnato's Central Company, whatever it cost. With the price war having seen diamond prices fall to an unprecedented eighteen shillings per carat, he now watched Central Company shares soar from £14 to £49 as he and his adversary scrambled for ever more stock. Again Beit came to his friend's rescue, partly by subscribing £250 000 himself, and partly by selling the concept of amalgamation to international financiers. Without that sort of support, who knows how differently South African history would have unfolded? In the end Rhodes had bought 60 per cent of the stock of the Central Company, putting him in an unassailable position. In March 1888 Barnato capitulated.

Together with Beit and Charles Rudd – Rhodes's first partner in the De Beers diggings – and other De Beers shareholders, Rhodes acquired the Central Company, effectively gaining control of the diamond fields. Barnato, on the other hand, became the owner of a substantial block of shares in the company that now owned both mines. It would soon be reconstituted as De Beers Consolidated Mines Limited (DBCM).

Cecil Rhodes, the Oxford gentleman, would continue to best the ex-showman from Whitechapel, Barney Barnato. Completely against his better judgement and after a long night of haggling, Barnato was persuaded to agree that profits from the De Beers mines would be used in furthering Rhodes's dreams of bringing the whole of Africa under the Union Jack.

If the agreement was hammered out in a simple corrugated-iron cottage with no more than the essential furnishings, which was how Rhodes chose to live, dry, dusty Kimberley had become, in some respects at least, one of the grandest towns anywhere. It had recently been connected to Cape Town by rail and was the first town in the world to boast electric street lights. One visitor went so far as to reckon that the Kimberley Club could claim 'more millionaires to the square foot than any other place in the world'.[5] This was only 11 years after Schalk van Niekerk's chance acquisition of a 'dull rounded water-worn pebble' on a remote farm on the very edge of the Empire.

The Kimberley Club and St Mary's Cathedral in 1884.

For a short time, there was relative calm. The Central Company went into voluntary liquidation and its assets were bought by DBCM for £5 338 650. The cheque that sealed the deal, signed on 18 July 1889 – at that time the largest ever written – is mounted on the panelled wall of the De Beers boardroom in Stockdale Street, Kimberley. It is a piece of paper you would not look at twice, its small size utterly belying its far-reaching historical significance. The final consolidation of the two big mines paved the way for the acquisition of the 'poorer mines', Bultfontein and Dutoitspan, and, still in the future, another pipe, Wesselton (originally the Premier), which would be brought into the fold, kicking and squealing, in 1891.

There were other dry diggings, too, between Kimberley and Colesberg, that would in the fullness of time join the Kimberley mines as part of the De Beers portfolio. Jagersfontein, in its early years owned by a group led by Barnato's nephew, Solly Joel, vies with the Kimberley Mine, now better known as the 'Big Hole', in two respects: as being the first kimberlite discovered and as the world's biggest hand-dug hole. Either way, it was already known as a source of diamonds far from any big river soon after the discovery of the Star of South Africa. Like the Big Hole, it too has a massive vertically sided pit to show, with a pond of green water flooring it. What is more important is that the Jagersfontein Mine has produced some of the world's biggest and most celebrated gems: the 995-carat Excelsior, cut into 10 gems between 13 and 70 carats, and the 650.8-carat Reitz, which was cut into a 245-carat cushion-shaped brilliant and renamed the Jubilee to commemorate the diamond jubilee of Queen Victoria's coronation in 1897. The lower-grade Koffiefontein Mine, also outside the De Beers stable until 1889, distinguished itself mainly in producing gems of exceptional clarity and colour.

The cushion-shaped 245-carat Jubilee Diamond, cut from the 650-carat Reitz (named after President Reitz of the Free State) in 1897, Queen Victoria's diamond jubilee year.

With his battle for control of the Kimberley mines won, Rhodes found himself free at last to pursue his ulterior and, by now, much bigger dream, the further colonisation of Africa. With his coup pulled off, there were good people to leave in charge of the day-to-day management of De Beers, while he threw himself with all his boundless energy into politics. To see the result of Rhodes's tireless scheming and manipulation in its perspective, the Cape's Chief Justice declared in a case put before the Supreme Court as the amalgamation process was reaching its climax that 'The powers of the Company [De Beers] are as extensive as those of any Company that ever existed.'

As important as the amalgamation had been in providing Rhodes with the might he needed to pursue his long-term objectives, it secured the price stability for diamond sales that had until then been tenuous at best. With most of the production in the hands of a single company there could be none of the undercutting that had so recently brought diamond prices to a level below the cost of producing them. Now De Beers could carefully regulate supply, a process so essential to ensure that the supply–demand balance remained tipped in its favour.

Although Rhodes had the funding to put his imperial dreams into practice, he had no political base. He set about applying his not inconsiderable persuasiveness and far greater determination to achieving that end, moving from MP for Barkly West to Prime Minister of the Cape in 1890. From Cape Town, with a lot of help from his older, wiser and less fiercely

Rhodes Memorial

Cecil Rhodes's legacy

Over a hundred years later, how is Rhodes remembered? In the countries that once proudly carried his name, Southern and Northern Rhodesia, now Zimbabwe and Zambia, history books probably demonise him if he appears at all. In Oxford he lives on. People from across the globe are entered there as Rhodes Scholars: for many it will be the proudest line in their CVs. That scholarship links names as diverse as Edwin Hubble, Kris Kristofferson and Bill Clinton; prime ministers, governors-general, attorneys-general and Nobel Prize winners – pillars of history launched on their careers by Rhodes Scholarships.

Rhodes built a grand gabled house, Groote Schuur, designed by Herbert Baker, who became the major colonial architect of his time, and left it to the nation together with great tracts of land on the slopes of Table Mountain. Here were established the world-famous Kirstenbosch Botanical Garden and the University of Cape Town, which stands proudly looking to the interior – towards Africa – as Rhodes did.

It is a measure of Rhodes's love of his adopted homeland that he kept the property's original name, Groote Schuur, which literally means 'big shed'. It is anything but, as every prime minister after Union would testify. Groote Schuur, once the official residence, is mainly used now for state occasions and charity events, and is open to the public on certain days. The President and other top government officials live elsewhere on the extensive Groote Schuur estate, verdant heart of the Rhodes bequest.

On farms in the Groot Drakenstein valley near Stellenbosch after the vines had been devastated by phylloxera in the late 1890s, Rhodes, together with Harry Pickstone introduced the large-scale growing of deciduous fruit for export in the Cape, now one of the province's major industries.

All these – and much more – are the legacy of South African diamonds as personified by Cecil John Rhodes.

ambitious partner, Charles Rudd, he would step over the Transvaal Republic, across the Limpopo, and into the uncharted territory to the north, where legends of fabulous goldfields were secondary to his vision of the northern tribes standing proud under the Union Jack.

It was probably more in line with Rhodes's dreams to have a country named after him than a diamond. From his grave, though, he would have felt a twinge of envy for Thomas Cullinan, who gave his name to a stone far bigger than any ever found, of 3106 carats, as big as a man's fist. By the time Cullinan's name first sounded on Transvalers' lips, with 1902 drawing to a close, Rhodes had been dead for nine months. The tubercular teenager who had been sent to South Africa because of its healthier climate, and who had never enjoyed robust health, died of a heart condition in March 1902, aged 49.

In Pretoria, Thomas Cullinan would have read about Rhodes's death but may not have given much attention to it. He was at that time in the closing stages of an arduous negotiation to buy Elandsfontein, the farm of the Prinsloo family east of Pretoria, where he had found a diamond during clandestine prospecting four years earlier. To raise the £52 000 Cullinan needed to buy the farm and establish a mining operation, he had sold shares in a company he called the Premier Transvaal Diamond Company. Confident that a mine so far from Kimberley did not constitute a threat to its monopoly, De Beers declined to involve itself. When the funds from the initial raising dwindled to nothing, and with a few diamonds having been won from the primitive operation, Cullinan found two young brothers, Bernard and Ernest Oppenheimer, who were prepared to invest in the new mine. As a dealer in the family trading business in Kimberley, and with a passion for diamonds, Ernest, in particular,

The Premier Mine (now the Cullinan Mine) in 1927.

liked what he saw. His hunch would stand him in good stead, good enough to launch him on his singular dynastic path.

Within three years the production from the Premier Mine rose from 750 000 carats to 1 890 000 carats annually.[6] And it was not the scores of exquisite, flawless white stones and, early in 1905, the giant Cullinan diamond that had De Beers directors shaken, but the volume: nearly 2 million carats, from a pipe that, at 32 hectares, was as big as the five Kimberley pipes together. This was more than enough to undermine in a very powerful way the cozy monopoly that Rhodes had created. In time it would lead to a new order in diamonds. The decision by De Beers management not to invest in the Premier Mine undoubtedly marked a turning point in the history of diamonds.

In the ten months from September 1904 to June 1905, four diamonds bigger than 300 carats were found at Premier, including the Cullinan. With that flow of diamonds to dispose of, Premier took the inevitable step of forming its own marketing organisation in London, under the direct control of its board.[7] Single-channel marketing had been smashed. With a foot already in the door, Oppenheimer kept himself close to developments.

A hand full of diamond:
Gladys Snaddon (née Brown) remembers

'One day before I had started school my father rushed down the hill calling my mother to come at once with the three children. In starched pinafores – how these were produced under those difficult circumstances I do not know – but presently all of the few families working on the mine were assembled. McHardy (the general manager of the mine) said that a large diamond had been found, of great value. The mine employees [were] rugged sunburned men in thin shirts and pants, heavy boots, wide felt hats.

The announcement meant little to me, but I was amazed to find my father at my side urging me forward to the long sloping table under [the] zinc roof, supported by poles. M the manager placed a tiny stone in my left hand and a very large one in my right. My father supported my hand as I held the Cullinan Diamond in my rather small paw. The discovery of this gem created great interest around the world. The biggest diamond in the world! Several times I have had the pleasure of visiting the Tower in London where the Royal Crown and Orbs reflect back to me, reminding me of a very hot day in the Transvaal when it was put into my [hand] to hold – a little freckled maid of five …'

For a while all went well enough for both Premier and De Beers. Then, as competition mounted and Europe slid towards war, share prices of both companies plummeted. In the end it was the more experienced, cannier player that won the day, De Beers buying control of their rival as war broke out in 1914. Not far behind the scenes was Ernest Oppenheimer.

As the diamond boom died and the world mining community's attention shifted from precious stones to precious metal – gold – only a few hundred kilometres northeast of where South African mining had begun in earnest two and a half short decades before, Oppenheimer dug himself in. Not only did he have an uncanny sixth sense guiding him, but he was, as a strategist, every bit Rhodes's equal. With the backing of New York's J.P. Morgan Bank, he set up the Anglo American Corporation in Johannesburg to consolidate and expand his gold-mining activities on the Witwatersrand. But his first love would always be diamonds.

In 1919 De Beers suffered its second major setback. Like the first, this was a direct consequence of having lost the visionary that had led it to the pinnacle of prosperity. The company could not believe that the Germans would sell their recently discovered diamond mines near Lüderitz on the coast of South West Africa. These had already been expropriated by the South African government after the conquest of that territory, but in an extraordinary move De Beers forbore to take up the South African offer of involvement. In Pretoria the Prime Minister, Louis Botha, had no option but to accede to the request by Ernest Oppenheimer and his brother Louis that they be allowed to buy all the German mines, mines which between them were producing 18 per cent of the world supply. In the post-war world of diamond mining, the fish the Oppenheimers had landed was no minnow.

Three other parts of the world were soon added to the Oppenheimers' supply base: Angola, the Belgian Congo and West Africa, including, most importantly, Sierra Leone. Expelled from the syndicate which Rhodes had formed in London to market the De Beers production, in 1925 Oppenheimer, no stranger to diamond trading, formed his own. Not that he was alone. He had a powerful ally. Unlike his uncle Barney, Solly Joel, though the biggest single shareholder of De Beers stock, had no empire-building aspirations. All he wanted was the best possible outcome for diamonds, whether from his Jagersfontein or De Beers or Oppenheimer's sources.

For Ernest Oppenheimer things were moving fast. When diamonds were discovered at the beginning of 1927 in substantial quantity in Namaqualand, just south of the Orange River and not far from his South West African holdings, he realised his syndicate would have to absorb the massive injection of new production, whatever the cost. Had this new flow been allowed into an already depressed market, the result might have been catastrophic.

The Namaqualand threat materialised only months after another shock had shaken the diamond establishment. In the rich farming corner of the Western Transvaal, diggers were finding diamonds around the village of Lichtenburg in a profusion that turned the farmers

The two key 'discoverers' in Namaqualand, on the left the finder of the first diamond, Jack Carstens, on the right Hans Merensky.

greener than their summer maize stands. They were not coming from kimberlite pipes, nor from gravel beds of major rivers. To this day they remain a geological enigma. In the late 1920s all that mattered was that the flow of Lichtenburg stones into the market showed no sign of waning.

Oppenheimer knew that the same uncontrolled production had to be forestalled in Namaqualand at all costs. Failure to do so would tip the market from a precarious order into chaos. In Namaqualand the mining entrepreneur *extraordinaire* Hans Merensky had used his geological intelligence and what he had seen in the Lüderitz diamond diggings nearly twenty years earlier to his advantage. In the German fields he had noticed an infallible connection between rich concentrations of diamonds and fossil shells of the large warm-water oyster, *Ostrea prismatica*, which lay strewn over the ground, far more conspicuous than any diamond. Imagine his delight when he saw the oyster shells in Namaqualand: he was soon vindicated in assuming the same connection. Backed by another German doctor of geology with a taste for diamonds, Dr Reuning, he set about establishing himself as the controlling force in the Namaqualand diggings.

Foreseeing the inevitability of an uncontrolled free-for-all developing in that remote and desolate corner of the country unless urgent steps were taken, in February 1927 Merensky went to the highest powers in the land. As a seasoned operator not very far up the Atlantic coast, Sir Ernest Oppenheimer was there too. To make the point that what was happening in the northwest corner of the country should be taken seriously, Merensky started the proceedings by spreading two months of prospecting results on the table: 12 353 carats of diamonds. Sir Ernest must have gasped inwardly. This was nothing like the mixed bag coming out of Lichtenburg: every diamond on the polished wood surface was perfect, and the average stone size was far bigger than anything he had seen from his own workings south of Lüderitz.

Not only would he have to buy what he saw on the table and any future production, but, far more importantly, he would have to buy control of the H.M. Association, as Merensky's holding company was called, and without delay. His mind was working fast even before he left the sorting office of the Diamond Board of South West Africa, where they had met, that February morning.

Starting two months later, Oppenheimer and Solly Joel set about buying control of H.M. Association. The last payment, to Merensky's partner, G.A.E. Becker, was made on 12 July 1929, by which time they had acquired 74 per cent of the company, with Oppenheimer, in his own and Anglo American's capacities, holding the majority of the stock. Merensky sold his

Nearly a hundred years later: a De Beers operation in Namaqualand.

own shares in November 1928, for just over a million pounds. He retired from exploration to pursue his other love, farming, which he did in an idyllic part of the slopes of the Eastern Transvaal Drakensberg. Untiring, Sir Ernest Oppenheimer continued to entrench himself as the pivotal name in world diamonds.

The Namaqualand diamonds would repay the investment cost many times over, one mine in particular, at Kleinzee, producing regularly in excess of a million pounds' worth of diamonds annually. But by far the most important dividend from Oppenheimer's South African coastal diamonds venture lay across the Orange River in South West Africa. If they were south of the river, he argued, and at Lüderitz, far to the north, why not in the corner between the sea and the northern bank of the river? He soon had his answer. From small beginnings in 1928, the value of production by 1955 from the Consolidated Diamond Mines beach terrace deposit was £85 million, and production would continue at that level for the next 20 years.

All the while, as Oppenheimer widened his sphere of activity, he bought De Beers stock. Ten years after his first coup, his purchase of the Lüderitz mines, Sir Ernest Oppenheimer, now a pillar of the Anglican Church in Kimberley, won his biggest prize on 20 December 1929: the chair of De Beers Consolidated Mines. It could be said that he had ridden rough-

GARY RALFE

The internationally respected diamond industry guru Chaim Even-Zohar has said of Gary, 'The former managing director of De Beers fearlessly trod into uncharted territory, leaving giant footprints to guide generations to come'. He 'single-mindedly embraced legal and ethical compliance, transparency, good governance and accountability, … making every other corporate aim secondary to achieving those overriding objectives', tolerating 'no excuses, [making] no exceptions, and [allowing no] short-cuts or deviations'. In terms of industry statesmanship, Even-Zohar puts Gary on a par with Cecil Rhodes and Sir Ernest Oppenheimer, saying that he 'literally saved De Beers, together with all its stakeholders, from an inevitable non-future brought upon by its very own inaction, its complacency and its arrogance … acting out of a deep sense of loyalty to those he pledged to serve'. So, Gary's appearance here has nothing to do with the fact that he was my 'mentor' at boarding school in 1954 or that we still lunch together occasionally and enjoyably. The accolade is Chaim's.

shod over friends and family to get there. Perhaps it would be truer to say that what he got he wanted more than any of them. And that ambition drove him to analyse, to calculate and to strategise, endlessly, helped always by extraordinary intuition and by friends and family on the inside track of diamond dealing. What Rhodes had schemed and bullied to build, Oppenheimer won in a hard-fought game with never a call of 'foul'.

After his death Ernest's son Harry would follow in his father's footsteps, consolidating and expanding, opening huge new mines in South Africa and Botswana: Finsch and Venetia in South Africa, Orapa, Lethlekane and Jwaneng in Botswana. De Beers would take over the Mwadui Mine in Tanganyika when its Canadian founder and owner, Dr John Williamson, died. Harry and then his son, Nicky, would also establish De Beers in the frozen north of Canada.

For a hundred years, from 1888 to the late 1980s, De Beers ruled the world of diamonds almost single-handedly. Now, with the global village shrunk by technology, there are other major diamond producers around the world. In the important producing countries of

Harry Oppenheimer and son Nicky in earnest discussion.

Canada and Australia, De Beers was outflanked. Even in its own backyard, South Africa, it has sold off the Premier and Koffiefontein mines to a smaller company, Petra Diamonds, which has also bought the Mwadui Mine in Tanzania.

If De Beers' role in mining is less exclusive than it once was, that there is still a diamond industry at all is thanks to its marketing. With order books in the United States, Europe and Israel as full as ever and new markets burgeoning in China, India, Japan and the Gulf, an environment has been created that no longer needs single-channel marketing. That environment is the construct of a De Beers advertising strategy which has spanned many decades and cost many millions of dollars. The company still pursues the campaign relentlessly, to the advantage of anyone who owns a diamond or works in the industry. It has worked better than most of us realise.

Demand for diamonds is stronger than ever. The discovery of new mines has slowed to a snail's pace – a natural decline that not even De Beers could have engineered in our age of transparency. Supply, demand and prices balance themselves without any intervention. But we need to remember it was not always so. Without the fierce protectionism practised in those early days, who can say whether the industry would have survived?

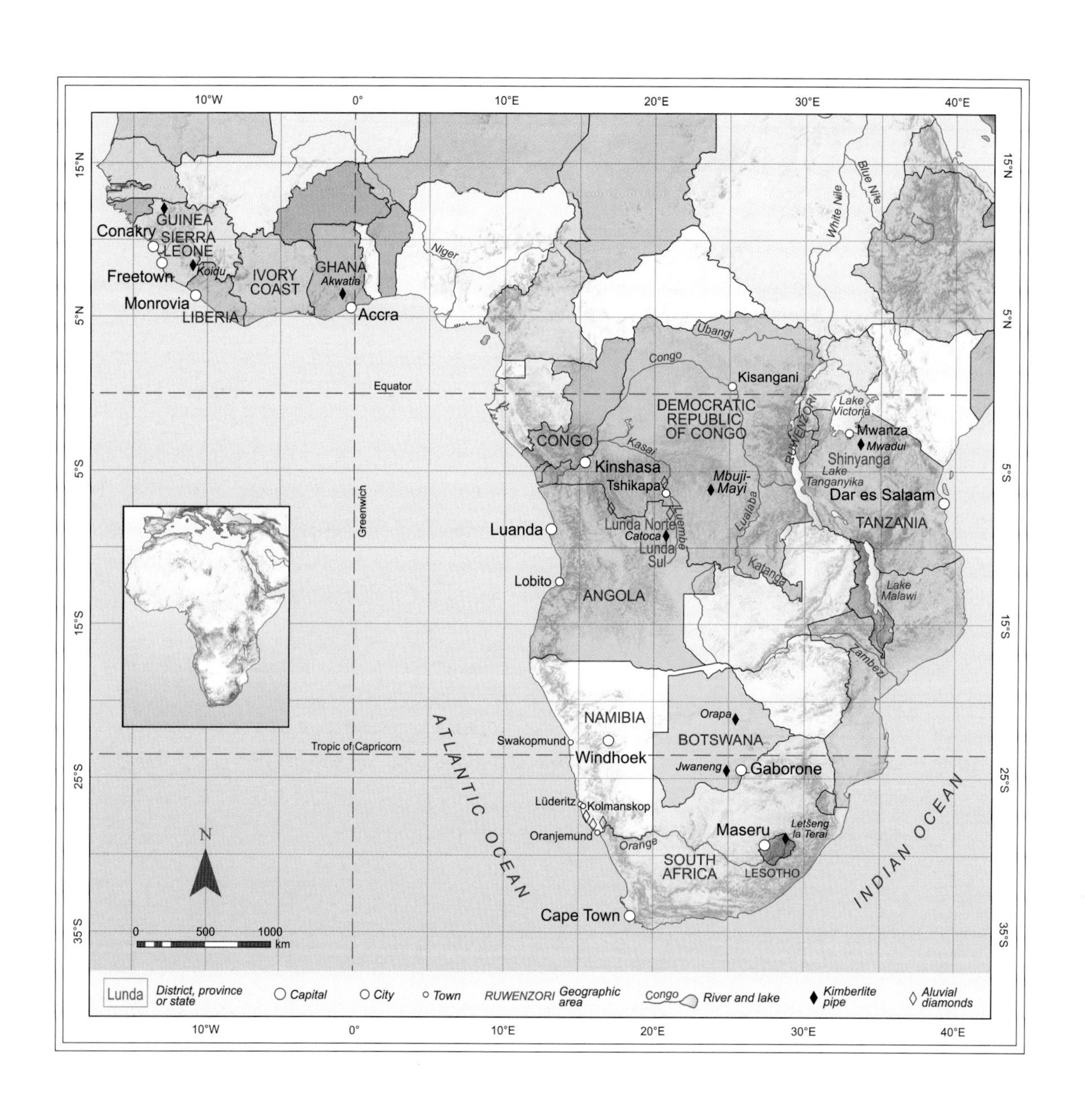

GUINEA
Conakry
SIERRA LEONE
Koidu
Freetown
Monrovia
LIBERIA
IVORY COAST
GHANA
Akwatia
Accra
Niger
Equator
Greenwich
Ubangi
Congo
Kisangani
DEMOCRATIC REPUBLIC OF CONGO
CONGO
Kasai
Kinshasa
Tshikapa
Mbuji-Mayi
Lunda Norte
Catoca
Lunda Sul
Luembe
Lualaba
Katanga
Luanda
Lobito
ANGOLA
RUWENZORI
White Nile
Blue Nile
Lake Victoria
Mwanza
Mwadui
Shinyanga
Lake Tanganyika
Dar es Salaam
TANZANIA
Lake Malawi
Zambezi
NAMIBIA
Swakopmund
Windhoek
Orapa
BOTSWANA
Jwaneng
Gaborone
Lüderitz
Kolmanskop
Oranjemund
Orange
Maseru
Letšeng la Terai
LESOTHO
SOUTH AFRICA
Cape Town
Tropic of Capricorn
ATLANTIC OCEAN
INDIAN OCEAN
N
0 500 1000 km
Lunda District, province or state
Capital
City
Town
RUWENZORI Geographic area
Congo River and lake
Kimberlite pipe
Aluvial diamonds

CHAPTER SEVEN

THE REST OF AFRICA

Once diamonds had been found in South Africa, it was natural to expect that they would occur elsewhere in Africa. It is a big continent, after all, and there was no reason to suppose they would be found only at its southern tip. Further north, however, new diamond discoveries did not follow as soon as might have been expected. This is because southern Africa – particularly South Africa – has evolved differently, climatically and geomorphologically, from the rest of Africa, not enormously but enough to have a direct effect on the distribution of diamonds. In the south of the continent diamonds could be found quite easily lying on the surface, but elsewhere not.

In fact it was to be forty years after the Eureka was found that the first finds were made in the Congo. After that there was a steady, if at times slow, addition of new members to the list of African producers. The sequence of discovery dates of the most important producers looks like this:

Congo	1907	Sierra Leone	1930
Namibia	1908	Tanzania	1940
Angola	1916	Lesotho	1952
Ghana	1919	Botswana	1969

As though to emphasise that the order has nothing to do with the quantity of diamonds in any one part of the continent, it is the first and last countries on the list that are, or have been, Africa's most prolific producers. This point leads to another, the randomness of the world of diamonds: the Congo produces an unusual preponderance of industrial stones, whereas Botswana has a disproportionately high percentage of gems.

The circumstances of discovery in each country are so distinctive that we shall discuss them separately. And because they occur within a context of evolving global politics and technology, we shall treat them chronologically in the order of discovery. Let it be said before going any further that there are a number of other countries, mainly in West Africa, that are minor producers, whose songs will have to remain unsung.

The story of the unfolding of Africa's diamond wealth needs to be prefaced by the reminder that in most, if not all, cases, the shiny hard stones were known to local travellers long before they came to the attention of their colonial overlords. The Africans – natives of the depths of forest and savannah for countless generations – were the true discoverers.

One last generalisation about the African diamond narrative, to bring us almost to the present, concerns the fascination of blood diamonds or, as they're more correctly known, conflict diamonds. While they are thought to be ubiquitous throughout Africa, in fact the 'blood diamonds' label applies strictly to only three of the countries for reasons that neatly link geology and history. The three are Sierra Leone, the Congo and Angola. The common denominator is the tremendous richness of their alluvial deposits of gems. In all of them there are extensive concentrations of diamonds near the surface and close to their sources, a hundred times more prolific than in the South African alluvial deposits. As a result, artisanal working can recover thousands of carats of highly marketable diamonds in a relatively short time.

For the reason why the super-rich diamond deposits of Sierra Leone, the Congo and Angola were bound to lead to conflict, one does not have to dig very deep. In the Congo and Angola a political vacuum was left after decolonisation, which was accomplished with inadequate preparation and in haste; in Sierra Leone the process took longer. Nature and politics abhor a vacuum. Diamonds – the most concentrated and easily transportable form of wealth there is – could buy guns and tanks, and they, in turn, could buy power, inevitably at an awful cost in human life. That is the simple fact behind conflict diamonds.

Democratic Republic of Congo

In the days when it was still the Belgian Congo it was realistic to think of it as a single country. Today, after decades of ethnic conflict and corrupt government, the national infrastructure has deteriorated to the point where geographically separate parts of the Congo exist practically in isolation, and normal travel from one to the other is only by air.

A hundred years ago the infrastructure across that huge country – Africa's third largest – was better than today, and administrative outposts were in regular contact with each other and with the capital, Leopoldville (now Kinshasa). A latex-yielding creeper had been discovered in the Congo basin and when, in 1888, John Boyd Dunlop patented the pneumatic tyre for the bicycle, there was an explosion in the demand for latex from the Congo. It overtook ivory as the main export, and King Leopold II prospered more than ever.

Leopoldville, too, was in constant communication with Belgium. We can suppose that in Antwerp, diamond capital of the world, there were those wondering whether they could not secure for themselves a supply from their king's own colony. It is no surprise, then, that the first African country where diamonds would be found that were not scattered over the

Deep in central Africa, getting around is easier for some than others.

surface was the Belgian Congo. And with the fabulous richness of some of their deposits, the stones, and their source, were bound to be discovered by the Belgians before too long, in fact just over twenty years after colonisation by Leopold II.

In November 1907, thirty years after Stanley's historic expedition down the Congo River, a mixed party of Belgian and American prospectors were prospecting in the southern part of Leopold's territory. They sieved and panned river gravels along the numerous main rivers and their tributaries, mainly, it seems, in the hope of finding gold. Their reception by the local people was less than friendly, and after four months of gruelling conditions they were preparing to turn north and back to civilisation.

During his sieving of gravels on the middle reaches of the Kasai River, the Belgian prospector Narcisse Janot found a very bright colourless stone, which was too small for either him or the party's geologist to identify with any certainty. It was carefully stowed away in a test-tube and duly forgotten about. Further north, they spent another eventful fifteen months prospecting before returning to Belgium.

It was only in September 1909 that the test-tube containing the unidentified stone was opened by a Belgian mining engineer. He was preparing, with extraordinary thoroughness, to lead a follow-up prospecting expedition to the Congo. This time there was a microscope to hand, and the stone was discovered – in what must have been a Eureka moment – to be a diamond. Brains were racked and notebooks scoured, and, in the end, the provenance of the stone was run to ground.

Janot did not return to the spot for nearly two years, and it was only in August 1911 that he found the Congo's second diamond, and the third and a lot more the same day. Within less than a month 242 stones had been found in the vicinity of where the Tshikapa River joins the Kasai, the latter one of the main tributaries of Africa's greatest river, the Congo. The Belgian Congo was now on the map as a diamond producer.

The timing was unfortunate, though. The discovery in 1908 of easily accessible and rich fields just outside a busy port in German South West Africa meant that by the time the Tshikapa area was in its first production the flood of South West African gems had already depressed the European markets. Then came the First World War. This was only a temporary setback: by 1916 some 50 000 carats were being produced, double the next year and 160 000 in 1918. In time primitive manual conveyance of river gravel and its sorting gradually gave way to conveyor belts and mechanical sorting. So extensive were the gravels around the settlement of Tshikapa that by 1957 there were 53 producing mines. This represented the peak of activity. In the ensuing years the number of operations diminished as did their scale, so that by now the area is to all intents and purposes worked out.

About 300 kilometres to the east, another bonanza area was turned up during the early exploration, in 1918. The Mbuji-Mayi deposits are so extraordinarily rich in terms of carats per tonne that the early diggers could scarcely believe their luck – until they came to sell the stones. They were almost unmarketable, even to industry, which in those days had limited need of diamonds for grinding, cutting and polishing. The Belgian-American mining company continued processing the surface material that contained the diamonds, pulling out the occasional gem and storing the small grey and black stones in milk cans, hoping for better days.

VERORDNUNGSBLATT

des Militärbefehlshabers in Belgien und Nordfrankreich

für

die besetzten Gebiete

Belgiens und Nordfrankreichs, herausgegeben vom Militärbefehlshaber (Militärverwaltungschef)

Erscheint nach Bedarf — Einzelpreis : 20 Rpf.

20. Ausgabe	ABDRUCK — auch auszugsweise — VERBOTEN Ausnahmen genehmigt der Militärverwaltungschef	Den 5. Nov. 1940

INHALT:

1. Verordnung über Massnahmen gegen Juden (Judenverordnung) vom 28. Oktober 1940. — 1. Verordening van 28 Oktober 1940, houdende maatregelen tegen de Joden. (Jodenverordening). — 1. Ordonnance en date du 28 octobre 1940, concernant les mesures contre les Juifs. (Ordonnance relative aux Juifs).

2. Verordnung über das Ausscheiden von Juden aus Aemtern und Stellungen, vom 28. Oktober 1940. — 2. Verordening van 28 Oktober 1940, betreffende het verwijderen van Joden uit ambten en betrekkingen. — 2. Ordonnance en date du 28 octobre 1940, concernant la cessation de l'exercice des fonctions et activités, exercé par les Juifs.

1.

Verordnung über Massnahmen gegen Juden (Judenverordnung) vom 28. Oktober 1940.

Auf Grund der mir vom Oberbefehlshaber des Heeres erteilten Ermächtigung verordne ich für Belgien folgendes :

I. ABSCHNITT

Begriffsbestimmung des Juden.

§ 1

(1) Jude ist, wer von mindestens drei der Rasse nach volljüdischen Grosselternteilen abstammt.

(2) Als Jude gilt, wer von zwei volljüdischen Grosselternteilen abstammt und

1. entweder im Zeitpunkt des Inkrafttretens dieser Verordnung der jüdischen Religionsgemeinschaft angehört oder danach in sie aufgenommen wird, oder

2. im Zeitpunkt des Inkrafttretens dieser Verordnung mit einem Juden verheiratet ist oder sich danach mit einem solchen verheiratet.

(3) In Zweifelsfällen gilt als Jude, wer der jüdischen Religionsgemeinschaft angehört oder angehört hat. Ein Grosselternteil gilt ohne weiteres als volljüdisch, wenn er der jüdischen Religionsgemeinschaft angehört hat.

The occupation of Belgium by Nazi Germany in the Second World War meant that Sir Ernest Oppenheimer, considered to be Jewish, was barred from serving on the board of Sibeka, the Belgian diamond company operating in the Congo.

At first it was assumed these were alluvial deposits, left by the rivers before they cut down to their present level, but as the miners excavated deeper into the central zones of concentration, they found the stones were not in gravel. They were in clayey material that became harder as the depth increased until it was clear that the diamonds had not been transported at all, but that this was the top of a group of six kimberlite craters. More were discovered nearby. Now that the Mbuji-Mayi cluster has shown that it is sustainable year after year, we see that the description of the area as 'one of the wonders of the world of diamonds' is well earned.

What of the diamonds accumulating by the milk pail? As if to lend credence to the saying that 'it's an ill wind turns none to good', the Second World War brought relief and by 1945 the pails had been emptied and the stones were selling at $1 a carat, a tenfold increase on the prices twenty years earlier. Still the stones continued to pour out of the pipes by the bucket-load. At the start of the war, in 1939, the Congo was producing 67 per cent of the world's production, and the great majority of it was coming from Mbuji-Mayi.

Since then, other alluvial deposits have been brought into production, on the Kwango River, which flows in from kimberlite-rich western Angola, and in the centre of the country from an area north of the major regional city of Kisangani (formerly Stanleyville). South of the town of Tshikapa, too, on the river of the same name, and on a number of parallel north-flowing rivers from Angola, there are alluvial workings, though the stones, all coming from Angolan kimberlites, are small this far north of their source. The main production is still from Mbuji-Mayi, and although the Congo has long since lost its pre-eminence, falling behind Russia, Botswana and Australia, the Mbuji-Mayi deposit has far outstripped any other kimberlite field in its total production.

Namibia

The contrast between the birth of the diamond industry in the Congo and German South West Africa – as Namibia was then called – is nothing short of dramatic. Consider how tentative was the start of diamond mining in the Congo, how nearly it hadn't happened at all when it did. The Namibian find, by contrast, was reminiscent of an express train, and the metaphor is in fact appropriate.

In his classic book *The Diamond Fields of Southern Africa*, Percy Wagner comments on the find eloquently and fittingly: 'Africa, proverbially the land of surprises since Pliny's famous *dictum* of 2000 years ago (*Ex Africa semper aliquid novi* – Always something new out of Africa), had this time excelled herself.' In April 1908 Zecharias Lewala, a 'Coloured' labourer from the Cape, was shovelling drift sand off the railway line at Kolmanskop, 15 kilometres inland from the port of Lüderitz. When told by a railway official, Herr Stauch, to look for diamonds, the clearing gang had laughed. August Stauch, recently arrived from Germany, was a romantic who was sure the sand hid untold treasure. Neither Lewala nor the gang foreman to whom he gave the pretty shiny stone he found that day in April could say it was a diamond, but when it was given to Stauch and he scratched his watch glass with the stone, he knew exactly what it was.

Early days in Namibia's Sperrgebiet, the forbidden diamond area.

Without a word to his friends, Stauch applied for a claim from the German Colonial Company of South West Africa, the DKG, over the area where the stone – and others soon after it – had been found. Stauch knew this was something much bigger than just a few small diamonds along the railway line. Like everyone in Lüderitz, he was familiar with the summer gales that raged off the wild Atlantic for days on end, when the sand blew from the coast, from southwest to northeast. The diamonds had been blown to where they'd been found. Stauch pegged more claims to the south and southwest of the discovery point, and resigned his job.

Still not having had any authoritative confirmation that the stones were diamonds, Stauch travelled to Swakopmund, where there was a minerals laboratory. As his luck would have it, a European gem expert visiting the laboratory quickly confirmed the stones were diamonds. Stauch hurried back to Lüderitz and claimed more ground. Together with two financial backers, he founded the Colonial Mining Company. The backers, we are told, returned to Germany before long, wealthy beyond their wildest dreams, never to return to the colony.

Stauch started his prospecting in the original claims straddling the railway line, in the 'streets' or sand-free valleys between the dunes, where he made a rich haul of very evenly sized diamonds, four or five to a carat. He may well have realised that the size uniformity had come about because of the better sorting capacity of wind compared with water.

The same scepticism shown by the railway workers before the discovery prevailed in the bars and streets of Lüderitz when the news of Stauch's mining success was put abroad. Slowly, optimism gained the upper hand and a rush into the desert got under way. Fred Cornell, in his narrative of endurance and hope *The Glamour of Prospecting*, described the types he encountered in the early days of the stampede: 'shady "company-promoters", bucket-shop experts, warned-off bookmakers and betting men ("brokers" they usually styled themselves), and sharpers of all sorts, on the look-out for prey in the shape of lucky diggers or discoverers … runaway ships' cooks, stewards, stokers and seamen.' It was reminiscent of Kimberley in the early days.

As regards the mining itself, it was of unparalleled simplicity. Cornell was there himself, so let him speak again: 'This loose deposit [sandy shingle], in common with the sand, is in many places heaped up into small, wave-like ripples by the action of the prevailing wind, and wherever the diamonds exist these little ridges are exceptionally rich in them. The method of searching for them was simply by crawling along almost flat upon the ground, and turning over the shallow layer with a knife point, though on some of the claims hand-washing in sea water was being attempted.'

While the grade was higher than in any deposit before or since, and the stones were of the very top quality, they remained small, the largest that year (1908) being 2½ carats. Small stones or not, their quality and the ease of recovery meant that the profitability of the mining was phenomenal. Records show that mining cost was 2 to 3 marks per carat, with sales value of stones 28½ marks per carat, and within a year the production was 500 carats per day.

The northern Sperrgebiet is characterised by small to very small diamonds.

Before long Stauch had a busy prospecting and mining operation going in the area just south of Lüderitz and Kolmanskop. He was able to leave on a more extended prospecting trip, to see what lay beyond the forbidding sand cordon not far to the south of Lüderitz. Together with a few labourers he set off with a recent associate, Professor Scheibe, director of the mineralogical department of the Royal Mining Academy of Berlin, who had come to the colony on a year's sabbatical to study a field of kimberlite pipes inland of Lüderitz. Twice they tried to cross the dune cordon. Twice they failed. The labourers were loath to try again. But Stauch was more determined than ever. Their third attempt successful, they reached a wide valley which Stauch named Idatal (Ida's Valley) after his wife.

Returning to the camp from the coast on their second day there, Stauch jokingly admonished the labourer who had been sent for driftwood for the fire, saying he should be looking for diamonds, not wood. Ever dutiful, the man dropped his load and got to his knees. Searching on hands and knees was, after all, standard exploration practice. Stauch heard him give a yelp and saw that already his hands were full of diamonds, and that he was now stuffing them in his mouth to leave a hand free. Dismounting, Stauch called to the professor and joined the labourer on his knees. The evening fire could wait. Between them in a few hours in that single place they collected handfuls of diamonds. Before the day was over the place had been nicknamed Märchental (Fairytale Valley).

That night the men were glad they weren't alone.[1] They needed to hear a voice that wasn't in their head, saying it wasn't all a dream. While they waited for the evening onshore wind to gather strength they watched the moon rise, and no wind came. It was eerily still and clear. Sleep was out of the question. Minds as active as they'd ever been, they got out of their bedding rolls and walked back to the place. Little eyes blinked at them. They shook their heads and walked back. The moon was low in the sky when finally they went to sleep.

The next day Stauch and Scheibe sought out the best-looking areas and pegged some ten claims. It wasn't long before the news had filtered back to Lüderitz and the Pomona rush was on. In twenty months of unmechanised mining – mostly hand-picking off the surface – a million carats were mined. That is 200 kilograms, or ten of the suitcases you check in at the airport, watching the scale reading and hoping your bathroom scale hadn't lied.

But 'mining' didn't get under way immediately. It had to wait for the complex issue of the legality of the claims of Stauch, Scheibe and others to be resolved. Eventually, after letters had travelled between Lüderitz, Berlin and London, and months of costly litigation, the Pomona Diamond Company was formed in Berlin, with Stauch holding 49.5 per cent, Scheibe's company 21 per cent and Cape Town parties the rest. By 1914 the company was recovering 50 000 carats per month. The area was – and remains to this day – unique in so far as concentration of diamonds is concerned. While rich pockets of diamonds would be found further to the south in ensuing decades, nothing occurred on this scale.

Bigger stones were being found, too, up to 10 carats, whereas those nearer Kolmanskop and Lüderitz were only very rarely up to 2 carats. There was every reason to push southwards and see if the trend continued. But even at Pomona logistics had become a major concern, especially fresh water. Sea water could be used for washing gravel, and at the end of a day spent so close to the ground a dive into the breakers would wash the grime off, but drinking and cooking water for both men and pack animals had to be as carefully rationed as other provisions. And the further from Lüderitz the greater the problem. Bogenfels was to be the limit of the southward push, beyond which was to remain *terra incognita* – to the diggers at least.

Bogenfels is a place that is impossible to miss, and equally impossible to forget once you've seen it. It is a massive buttress jutting out from the coastline, with nearly its entire core eaten out by waves and wind over millions of years, so that it stands as an arch (*bogenfels* means 'arch rock'), one leg on the shore, the other planted defiantly in the waves. If the wind is not

Bogenfels ('arch rock'), a spectacular landmark on the Sperrgebiet coast.

howling and you have a head for heights, you can walk to the peak of the arch across a bridge suspended over the surging surf. Inland and within sight of the arch, when I was there in 1964, heaps of small pebbles from which the diamonds had been washed could still be seen.

South of Bogenfels the diamonds on the surface were few and far between and were left to lie there. Two prospectors working for the German Diamond Company who did cross the wastes into the unknown were Georg Klinghardt and the geologist Dr Ernst Reuning, great pioneers both of them. In 1910 the two crossed to the Orange River, enduring untold privations and hardship in an area as devoid of water as it was of any maps. Men whose very survival depended on taking in every aspect of their surroundings without knowing, unquestionably saw far more than we, with a focus that is so different, would today. Along the Orange River they noted old gravel terraces on both sides of the river. They met up with a farmer near the Orange River and Reuning arranged for him to dig some prospecting pits, which he would return to inspect. If, as he suspected, the diamonds along the coast near Lüderitz had been brought to the Atlantic with its north-flowing Benguela current by the Orange River, these gravels must carry diamonds.

In the meantime Reuning and Klinghardt travelled upriver into the mountain country, where the river flows mostly through deep gorges. At Sendelingsdrift (Missionary's Drift), deep in the mountains, Reuning found a messenger waiting for him with a note that he was urgently awaited at Lüderitz. He never inspected the pits he had had dug.

At Sendelingsdrift Reuning saw more evidence of old terraces on the south side of the river, in South African territory. Though he would return to pit these, too, he would have to leave without the answer he sought, perplexed and frustrated at not being able to dig deeper by hand. Fifty years later the visionary Baxter Brown would deepen one of his pits and find diamonds less than a metre below the bottom of the old pit. The major mine developed there would be known as the Reuning Mine.

Dr Hans Merensky points out a diamond in a rock sample from the oyster line.

Another name associated with the Lüderitz diamond fields and immortalised in mining history is that of Dr Hans Merensky, though he, too, is best remembered in the world of diamonds for his activity south of the Orange River. Yet he might not have achieved his remarkable success in Namaqualand had it not been for the time he spent in the field between Lüderitz and Elizabeth Bay, 30 kilometres south of it.

In 1908 Merensky was sent by a very concerned South African mining community to evaluate the sustainability of the diamond field being opened up around Lüderitz. Alarming news was coming in from the German outpost in the desert. Being of German descent, Merensky would quickly gain the confidence of the local prospectors, and the respect he had earned as a consulting geologist in South Africa ensured a reliable report. It was hoped he would return to Johannesburg with tales of an isolated occurrence with limited potential. If not, the damage to the diamond industry worldwide, and South Africa in particular, could be devastating. The Lüderitz stones were all going via Germany into the cutting and polishing houses in Europe, and if these quantities were such as to drastically expand overall production, the downward effect on the price of all rough sold could put the whole global industry at risk. Already De Beers and Thomas Cullinan's Premier Mine had been able to agree to limit the release of South African stones to Antwerp and Amsterdam. The South West African development could spell disaster.

By 1914, when mining had been substantially scaled up from the surface hand-picking, the supply overhang of the market started to manifest itself to everyone's discomfort. It was time for more direct action. The diamond expert Ernest Oppenheimer, then unattached to De Beers, accompanied a high-level De Beers delegation to Lüderitz, where Stauch and others proudly showed off their massive new plants and their plans. A few months later senior representatives of the main South African and South West African producers gathered in London. A system of quotas was agreed on for the four main producers, supplying into a pool of maximum value £12.5 million. The South West African producers' quota was to be 21 per cent of the total and De Beers 48.5 per cent, with Premier 19.5 per cent and Jagersfontein 11 per cent. Less than two months after the London meeting, on 4 August 1914, war was declared. In southern Africa neighbours of long standing were no longer friends, and hostilities were inevitable. South African forces invaded South West Africa, forcing the Germans ever northward until a treaty was signed in July 1915. Only three months later, small-scale mining was resumed around Lüderitz.

Although in terms of the treaty all South West African government property was to have been disclosed to the South African authorities, a parcel of diamonds was spirited away and hidden in a military kitbox in an antbear burrow on the veld in the far northeast of the country. The South African intelligence chief, suspecting there must be diamonds stored where none had been disclosed, used the threat of cancellation of the treaty to force their exposure. Out of an animal digging like thousands of others came 75 000 carats of top quality diamonds, including a 40-carat lemon yellow stone, worth in today's currency many millions of dollars. Because the stones were not, however one looked at it, government property, there had, in fact, been no obligation to disclose them. Nonetheless, the cache was seized and held by the occupying forces, to be sold after the war, with 50 per cent of the proceeds paid out to the companies concerned in a gesture of questionable magnanimity. It only goes to illustrate the truth that 'to the victor belong the spoils' or, in this case, half of them.

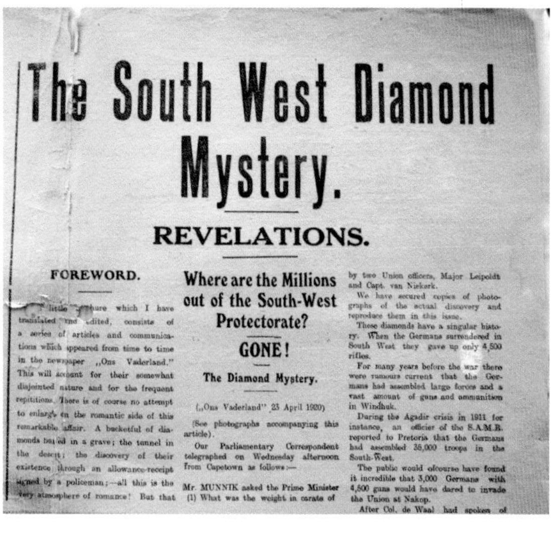

The South West Diamond Mystery.

REVELATIONS.

FOREWORD.

little [illegible] which I have translated and edited, consists of a series of articles and communications which appeared from time to time in the newspaper „Ons Vaderland." This will account for their somewhat disjointed nature and for the frequent repititions. There is of course no attempt to enlarge on the romantic side of this remarkable affair. A bucketful of diamonds buried in a grave; the tunnel in the desert; the discovery of their existence through an allowance-receipt signed by a policeman;—all this is the very atmosphere of romance! But that

Where are the Millions out of the South-West Protectorate?

GONE!

The Diamond Mystery.

(„Ons Vaderland" 23 April 1920)

(See photographs accompanying this article).

Our Parliamentary Correspondent telegraphed on Wednesday afternoon from Capetown as follows:—

Mr. MUNNIK asked the Prime Minister (1) What was the weight in carats of

by two Union officers, Major Leipoldt and Capt. van Niekerk.

We have secured copies of photographs of the actual discovery and reproduce them in this issue.

These diamonds have a singular history. When the Germans surrendered in South West they gave up only 4,500 rifles.

For many years before the war there were rumours current that the Germans had assembled large forces and a vast amount of guns and ammunition in Windhuk.

During the Agadir crisis in 1911 for instance, an officier of the S.A.M.R. reported to Pretoria that the Germans had assembled 35,000 troops in the South-West.

The public would ofcourse have found it incredible that 3,000 Germans with 4,500 guns would have dared to invade the Union at Nakop.

After Col. de Waal had spoken of

This 1920 newspaper cutting alleges collusion between the Smuts-led South African government and the 'capitalists', namely Oppenheimer's Anglo American, in the disposal of diamonds seized after the First World War.

After the 1913–14 slump in prices, and mainly because of curtailed production in both South Africa and South West Africa during the early war years, the market started to recover by 1916. A year later the Lüderitz miners were producing at about a third of their pre-war rates and, thanks to improved prices, were showing a profit again. Yet the German miners must have asked themselves whether they would ever again see the profitability of those first years. During the slump, and then the war, they had seen the power of the entrenched players in the diamond world, the Diamond Syndicate, De Beers and others, prominently in evidence. Just before the war a rapprochement of sorts had introduced the various powers in the diamond world to each other. Now the question on everyone's mind was what sort of diamond landscape would emerge once the dust had settled.

Fearful of expropriation of his mines, Stauch had been persuaded that he should talk to any South African interests who approached with a view to amalgamation. It was generally assumed this must be De Beers, so there was general amazement when it gradually emerged that Ernest Oppenheimer and his Anglo American company, with full South African government backing, were the principal in forming the Consolidated Diamond Mines (CDM) of South West Africa. Stauch and the other producers would receive £3.5 million, half in cash and half in CDM shares. And De Beers could only watch from the sidelines.

The ghost of Cecil Rhodes, with his entrepreneurial flair, was dead, but in his place as the new giant of the South African mining industry emerged Ernest Oppenheimer. He had second-guessed the Stauch sentiment and prepared his discreet onslaught carefully. For its part De Beers had waited for the German mines to fall into its lap, like a ripe fruit. Even if they didn't, these were shallow alluvial fields producing small stones that could never compete with De Beers' own high-grade kimberlite mines. Had the deposits Oppenheimer bought just consisted of the underground parts of what had been mined on the surface, the thinking would have been right. But what Oppenheimer had seen in the north, around Lüderitz, was merely the tip of the iceberg: the main body lay to the south, in the corner between the Orange River and the ocean.

A later picture of Sir Ernest Oppenheimer.

Was Oppenheimer party to Reuning's thinking about the importance of the Orange River in the evolution of the coastal diamonds? Had he joined the dots? Probably not. If he had, he would have tested the hypothesis sooner than actually happened. In fact it was only once the diamonds had been found south of the river, in 1926, in almost the same profusion as at Pomona, that the dots were joined and the most extensive alluvial diamond deposit ever to have been mined anywhere in the world was discovered. At a time it was described as the biggest earth-moving operation in the Southern Hemisphere.

It needs to be said that this was not a low-grade deposit made commercially viable by economies of scale, as is often the case with mega-mines. This one's grade was higher than anyone dared hope, not only in terms of carats per hundred tonnes, but as dollars per carat, too. It was simply a one-off phenomenon in its total value.

Hundreds of kilometres of pounding Orange River floodwaters, and millions of years along some of the wildest littoral any ocean could muster, had allowed only perfection to survive. Near the mouth, with the longshore Benguela current a far milder driver than a raging river, the biggest gems had settled out, leaving the lighter, more transportable stones to pursue their journey along the coast, the smaller the further. And at the end, the smallest would finish their marathon sometimes rolling, sometimes bouncing, over the dry ground, gale force winds at their back. The unfolding of this unique drama was what awaited the geologists of Ernest Oppenheimer's CDM.

The big man himself would see that wild coast only seldom, though. He was too busy putting together a diamond empire that would dwarf Cecil Rhodes's De Beers of Kimberley. He would reshape De Beers Consolidated Mines to include Cullinan's Premier Mine, Solly Joel's Jagersfontein Mine and, by means of his own company Anglo American, CDM. This accomplished, De Beers became synonymous with diamonds.

After South West Africa's independence in 1990, when it became Namibia, CDM, by now already directly incorporated into De Beers, invited the Namibian government into its ranks as an equal partner in a new company to be called Namdeb. But by 1994, when Namdeb was fully constituted, the raised beach deposits of CDM were nearing the end of their very productive life. Even the beaches, mined by constructing massive berms to keep the waves out, had limited life left.

Part of the land-based CDM mining operation north of the Orange River.

By that time, De Beers had had ample time to evaluate the feasibility of mining undersea diamonds. Not only are any diamonds reaching the coast from the Orange River today swept into the ocean, but there were times in the geological past when the coast itself was well below today's sea level. Diamonds of that time, too, were trapped in potholes and gulleys, where they lie today, well below the surface. These have been the focus of attention of De Beers Marine Namibia's offshore diamond recovery vessels in the last 25 years. The company now operates five highly specialised vessels to mine diamonds in water depths ranging between 90 and 140 metres. The grades are low and the mining conditions as extreme as any. It is safe to say that only De Beers could dare to tackle a project of this magnitude, so far beyond the normal realm of mining. Even then, they only mine offshore at times when the resulting production can be sold at top prices. It is a world few of us can imagine.

It is one thing to push the sea back with gargantuan machines and to mine out into the surf zone; quite another to brave some of the wildest water on the planet, with *terra firma* no more than a low blur on the eastern horizon. It would all the same be incorrect to aver that it was De Beers that made the leap. In fact it was a Texan adventurer, Sam Collins, whose experience in laying offshore pipelines made him uniquely qualified to explore the possibility of marine mining. Collins, starting in the 1960s, was the pioneer whom De Beers has to thank, or curse, for having shown them that not even diamonds deep below the foaming whitecaps were beyond its reach.

Angola

After the Orange, the next big river up the Atlantic coast is the Cunene, beyond it Angola. The diamond environments of the neighbouring countries could hardly be more different. Geographically and geologically the Angolan diamond field is a continuation of the field in its northern neighbour, the Democratic Republic of Congo. The two are one, divided by an international boundary. But politics being what it is, they have evolved quite differently.

European heads of state at the 1884 Berlin Conference, as the carving up of Africa gets under way.

To begin with, it was almost as if there was no political divide. A hundred years ago most people in either country had only a vague idea where the border was. And as diamond mining got under way in the Congo basin, it was quite a new border, only established a quarter-century earlier by the Berlin Conference. The Berlin Conference has to be one of the most remarkable events in world history, if for no other reason than its sheer audacity and arrogance. Ambassadors of fourteen European powers assembled in Chancellor Otto von Bismarck's residence, with maps of Africa, pencils, erasers and rulers; their mission to carve up the continent. In November the map had been blank except for the main rivers and spots along the coastline and a short way up some rivers which had already been claimed by the more acquisitive of the fourteen. By the end of February it was crisscrossed with lines, some ruler-drawn, some along rivers and watersheds; and the pieces of the new jigsaw were given colours according to their new owners. Small wonder one mightn't be sure if one was in Leopold's domain or Lisbon's.

In July 1912 it was established that diamonds found on two tributaries of the Luembe, itself a tributary of the Kasai River, were in Angola. A Portuguese mining company was set up, with substantial Belgian and American interests, as well as some French and British investment. The operator, La Forminière, was the company the Belgians had set up to mine at Tshikapa, and the diggers and supplies came from there. The Angolan diggings became, in effect, a satellite of those at Tshikapa, even though they were in another country.

This cross-border diamond-mining relationship ended when a road was built to the Angolan diamond mines from a newly constructed railway line 600 kilometres to the south.

Mining Kwango River gravels in far northern Angola.

This line linked the Katanga copper mines deep in the centre of Africa with the Angolan port of Lobito, 2000 kilometres to the west. Both the road and the train line were accomplishments of extraordinary vision and engineering, and personal courage and determination, considering the forest-clad virgin terrain to be covered and the number of major rivers to be bridged. As a result Lunda Norte, in Angola's northeast corner, was opened up, to become one of the celebrated diamond-mining regions of the world. Gradually the frontier of the region where diamonds were found was pushed westwards and southwards. Always the stones were alluvial. They were found on the edges of the rivers, and higher up, on terraces where the rivers had flowed in earlier times. Arguing that diamonds must have washed into the present-day rivers, the miners diverted them, making new channels for them to flow in so they could get to the bed of the natural rivers, where they found the expected concentrations of good gems, and more. The early pickings in Lunda Norte were rich.

But where were they coming from? Finding the kimberlite sources of the diamonds proved to be a far more challenging job than any of the geologists assigned to the task had envisaged. It took decades of resolute follow-up of systematic geological mapping and shrewd deduction before the first kimberlite was found, in 1952. The discovery came as a result of the recognition that the diamonds were intimately associated with a graben or rift-valley structure, where a major zone of crustal weakness had allowed access to kimberlite pipes. Anyone who does diamond exploration in Angola today has learned about the Lucapa Graben running southwest to northeast long before he or she lands in Luanda. Although the term is now used quite loosely to define a belt that runs right across the country, the Lucapa Graben, as originally defined, is only 12 kilometres wide. It is a key part of the much wider belt that forms a priority target for diamond explorers.

Angola has a world-class kimberlite mine, Catoca, in production and a host of other known kimberlites, some of which must become producers in the fullness of time. Exploration of diamond-bearing alluvials continues apace and these remain an important contributor to the national inventory. Like many countries today, Angola also beneficiates its production on a growing scale.

Ghana

In Ghana – the Gold Coast of colonial times – long before gold-bearing veins and 'reefs' were followed underground from the surface in treacherous tunnels, gold was won by panning sediment from the rivers. So although the official date of diamond discovery there is 1919, one can assume that diamonds were found by gold panners hundreds of years before that.

Ghana is the dark horse of the diamond world. In less than a hundred years it has produced an estimated 100 million carats. But its name does not go up on any winner's boards, because by value per carat it straggles home last. The diamonds are tiny, at 25 to the carat not much bigger than a pinhead. At that size, although they are well crystallised, not even the most skilled craftsmen can cut and polish any but a few of them, and their value as industrial boart is below par. The 1971 *International Diamond Annual* tells us that 'the boart market in Belgium always tends to weaken when goods from Ghana are offered'. Consequently the Selection Trust company, which had for quite a sustained spell enjoyed great success in diamond mining in nearby Sierra Leone, was unable to justify maintaining its operation in Ghana despite the substantial economies of scale the enormous deposits offered.

Following independence, diamond mining was nationalised by the formation of Ghana Consolidated Diamonds, which is the biggest single producer, mainly from the Akwatia area in the Birim valley, northwest of the capital of Accra. Around 80 per cent of the declining national production (900 000 carats in 2003, down from an annual maximum of over 2 million carats) is from small workers, and smuggling of diamonds to adjacent Côte d'Ivoire and Liberia presents Ghanaian officials with continuing headaches.

The early geologists in Ghana were faced with a different set of headaches. Where had the diamonds come from? As there were none of the indicator minerals, like pyrope garnet and ilmenite, that often accompany alluvial diamonds, it was assumed that, like the Brazilian diamonds, they had been recycled from older conglomerates. But unlike the Witwatersrand conglomerates, which contain some of the oldest diamonds known, the Birimian gold-bearing conglomerates, which were the logical source, contained no diamonds. They were not coming from palaeo-river gravels.

Then, in 1943, came a breakthrough. A Geological Survey geologist found diamonds in a very ancient metamorphosed rock which he described as a phyllite. Phyllites are normally thought of as fine-grained meta-sedimentary rocks. It was over 60 years before a definitive study by a master's student, Dylan G. Canales, finally solved the mystery of how a huge resource of diamonds could be hosted by a meta-sediment so unlike the diamond-bearing gravels of Brazil, so unlike practically any other diamond host-rock. His clue came from what at first seems an unlikely source, until we reconstruct Gondwana.

Diamonds from a very peculiar rock-type had been described from French Guiana in 1999. They had been found *in situ* in a rock-type described as 'volcaniclastic komatiite'. Volcaniclastic is the term used to describe sediments that have been carried and worked by water (or wind) directly from a volcano; and komatiites are rare mantle-derived volcanic rocks, which are low in silica, potassium and aluminium, and high, sometimes extremely high, in magnesium. Komatiites are rare; volcaniclastic komatiites extremely so.

The research Canales did on the Ghanaian rocks showed that he was dealing with a very similar rock-type. Reassembling Africa and South America as they were juxtaposed before drifting separated them makes it very clear that what Canales found is exactly what one might expect. The corollary is that the Ghanaian and French Guiana diamond fields are as good a piece of evidence as one could ask for if one tries to prove that the continents had been joined and then drifted apart. The rock-type in the particular geological setting in which it is found is remarkable. When two examples of it on opposite sides of the ocean both constitute a unique host to diamond deposits, one really has no choice: one has to join them.

To end with a quote from the concluding section of Canales's thesis: 'Akwatia may be representative of a new and potentially significant type of diamond occurrence. Akwatia is a world-class diamond occurrence.' Let us not dismiss Ghana as a dead horse yet.

Sierra Leone

A thousand kilometres further along the African coast, where the bulge is nearing its westerly turning point in the Atlantic, lies Sierra Leone. Like Ghana, it is crossed by wide rivers that drain the interior highlands, some of which, together with a few tributaries, have carried diamonds along their courses.

Unlike Ghana, Sierra Leone has consistently produced a high percentage of beautiful gems, including some of the world's largest, like the 969-carat Star of Sierra Leone, the

Artisanal diamond mining in Sierra Leone.

770-carat Woyie River, and another two unnamed stones bigger than 500 carats. Reports tell of a 1400-carat stone that found its way from Sierra Leone to Antwerp via the Congo not long ago, but confirmation remains elusive. Even without it, Sierra Leone ranks next behind South Africa in the world list of large diamond producers. There can be few better illustrations of the fickleness of mother Earth than the geographic proximity of stellar Sierra Leone and humble 25-stones-per-carat Ghana.

In one important respect the Sierra Leone diamond environment is geologically more closely analogous with South Africa than the other big diamond producers in Africa. In Botswana, Tanzania, Angola and the Congo there are kimberlite pipes that have barely been eroded, if at all. The crater facies, as geologists call the uppermost part of the pipe, has been preserved. In Sierra Leone, on the other hand, when the first kimberlite was found, after the discovery of alluvial diamonds in 1930, it was quickly established that the source of the material was a dyke.

More dykes were found and 'blows' on dykes, with true pipes the exception. In South Africa some decades before 1930 the unravelling of the kimberlite mystery had shown that dykes represent the deepest part of the kimberlite body, and that only as the gaseous magma approached the surface was it able to open up a carrot-shaped chimney for itself. Dykes are common in South Africa because of where the kimberlites intruded in the east-central part of the subcontinent. There the uplift was most extreme, with over a thousand metres of the uppermost formations stripped off, including the tops of the kimberlite pipes. And although the pictures in one's mind's eye of Kimberley and Sierra Leone could hardly be more different, they share a history of repeated uplift and erosion. At around 300 metres, the diamond fields in eastern Sierra Leone, not far from the border with Liberia and Guinea, are not as high as the interior plateau of South Africa (1100 metres), but they were certainly higher in the geological past than they are now.

Diamond mining has occurred along the main rivers in the east, the Sewa and various tributaries in the district of Kono, as well as on dykes and in the country's main kimberlite pipe at Koidu. Because of the size and quality of the diamonds from Sierra Leone and growing political stability in the country, it remains high on diamond exploration companies' list of priorities.

For most people diamonds in Sierra Leone mean only one thing: a couple of hours of nail-biting drama and extreme violence with Leonardo DiCaprio doing a passable imitation of a South African accent. While the Congo's diamond-funded conflict was too deep in Africa to draw much press and the similarly financed civil war in Angola was principally of interest to South Africans because it was too close for comfort, Sierra Leone and adjacent Liberia were unquestionably the countries where, if not diamonds, at least the blood flowed most freely.

While for most of the 1930s and 1940s diamond mining was undertaken by the mining

An aerial view shows the scale of artisanal mining, much of it illegal.

company Selection Trust, and law and order prevailed, the end of the Second World War signalled the beginning of a new era. Soldiers returned with eyes opened to a life they had never dreamt of. Within ten years of the end of the war it is estimated that 30 000 people were illegally involved in the diamond trade. Selection Trust, as well as national administrators, saw that the rush could not be stemmed. The company was persuaded to reduce its land holdings drastically and the diggers were required to buy licences in order to continue. Now the rush was to buy licences. But foreigners were excluded, and tens of thousands of natives of neighbouring countries returned home, mostly no richer than when they'd arrived. While the measures resorted to might have slowed the illicit trade but only for a while, policing the new laws proved impossible.

Lebanese traders now became a key factor. For many years they had been trading legally, selling foodstuffs and basic necessities, but the profits were meagre and times were hard. Now, with illicit diggers anxious to dispose of diamonds hastily, they seized the opportunity to improve their station in life, quickly and at no great risk. Before long the Lebanese community had cornered the illicit diamond market.

With police in the capital, Freetown, well trained to pick up suspicious dealings on the street, and with the port and airport closely watched, it was a place to be avoided. Monrovia, in adjacent, more permissive Liberia, was a much better bet and the Liberian capital became the centre of Sierra Leone's illicit diamond trade.

No less concerned at the trafficking than the Sierra Leone government, whose lost revenue was huge, and Selection Trust, which was powerless to stop the pillage and pilferage, was De Beers. The quantity of top-end diamonds reaching the market in a completely uncontrolled fashion threatened price stability and constituted a precedent which could not be allowed. In the short term it might benefit illicit diggers and diamantaires as well as a few cutters from Antwerp and Amsterdam, but for the industry as a whole it was a menacing spectre.

De Beers was not going to beat them, so they joined the Lebanese traders, in competition of course. In 1956 De Beers formed the Diamond Corporation of Sierra Leone (DCSL), which would set up offices in the main centres of activity in Kono, alongside the illicit diamantaires. The offices would be manned by diamond sorters – experts in valuing diamonds – from Charterhouse Street who would buy diamonds from the diggers. The DCSL was later expanded to include an independent dealer from Antwerp and four others who would set the prices, higher than the DCSL had been doing. The new organisation, the Government Diamond Office, proved such stiff competition for the Monrovia dealers that they increased the legal trade in rough by a factor of six in two years.

For decades the Sierra Leone diamond industry developed in an orderly way. Then in 1991 the Revolutionary United Front (RUF) entered the diamond region from Charles Taylor's Liberia to launch their reign of terror. They blazed a trail of execution, maiming, and recruitment of pre-teen soldiers until four years later they seemed set to blockade Freetown. With nowhere left to turn to, in June 1995 the military government called in the South African mercenary unit, Executive Outcomes. In a book of that title, the organisation's founder reported: 'Businesses closed down and people began to evacuate the city [Freetown] in anticipation of it being overrun by the RUF.' One of EO's objectives was control of the diamond fields. 'As the rebels controlled the diamond and rutile mines – and therefore had unlimited access to funds – it would have been folly to allow them continued use of these natural resources.' A terse entry a little later: 'The advance on Koidu started at 06:00 on the 25 June 1995.' Within a few days the rebels were out of Kono and a few weeks later had been pushed back to isolated enclaves along the Liberian border. The RUF were not done yet, though, threatening to invade Freetown again in 1999 and 2000. In fact it was only in January 2002 that the war could realistically be considered over.

It would be sad to end the history of this country on a low note. Instead we'll finish with a reminder of the largest alluvial diamond ever found, the 969.8-carat 'D' colour ('blue-white') Star of Sierra Leone. It was found in February 1972 and sold to Harry Winston later that year. He had the doyen of master-cutters, Lazare Kaplan, cut the stone, Lazare having already cut the very tricky 726-carat Jonker diamond for him with great success in 1936. Out of the Star came two huge gems, one a 53.96-carat pear-shaped cut, the other a 32.53-carat emerald-cut, as well as a group of six smaller stones which Harry Winston made into a piece he called the 'Star of Sierra Leone' brooch.

As a postscript we should note that, having thrown off the 'conflict diamonds' stigma and been welcomed into the Kimberley Process club, Sierra Leone's diamond revenue has grown by leaps and bounds. It only remains to hope that some of that growth will filter back to the people.

Tanzania

Across the continent, and across the equator, the Tanzanian diamond field is as different from the West African ones as you'd expect. It is unique, though, in one respect: whereas various names pepper the history of every other diamond-producing country in the world, in Tanzania there is only one, Dr John Thorburn Williamson.

The mammoth Williamson Mine in the northern Tanzanian savannah.

It is possibly unique for another reason: it looks, unless and until a new discovery is made, like the only payable kimberlite not only in the kimberlite field but on the whole Tanzanian Craton. In all other fields the fertile pipes occur in clusters, not singly. And as if it were not singular enough already, there is another conspicuous feature to the Mwadui or Williamson pipe that distinguishes it from others: its size. At 146 hectares, it is nearly 40 per cent bigger than the next biggest pipe, Botswana's Orapa.

Williamson was blessed with extraordinary motivation. At McGill University he changed his course from law to geology, not stopping until he had finished his doctorate, en route having received the highest marks for his bachelor's degree exams ever given in geology. The McGill alumnus Austin Bancroft, on a recruiting drive at his alma mater for the Anglo American Corporation working on the Northern Rhodesian (now Zambian) Copper Belt, recognised the young man's passion for exploration and wasted no time in signing him up.

Big company bureaucracy soon palled and Williamson left the giant Anglo American to join a junior company exploring for gold and diamonds in Tanganyika (now Tanzania). Williamson was posted in the Shinyanga area, 150 kilometres southeast of Mwanza, on Lake Victoria. The company had found some small kimberlites which, though diamond-bearing, were not payable. Though Williamson found more kimberlites, these too were not mining propositions and the company decided to move on. Not Williamson, though: he was convinced the area had not been thoroughly tested.

With no financial backing other than his own savings Williamson set out to find the kimberlite that had released the diamonds occasionally recovered from stream gravels around Shinyanga. What he lacked in outside funding he made up for with a passion that was irresistible and a conviction that those who lent him money and supplied provisions 'on tick' would be repaid in full. In March 1940 Williamson had reached the end of his tether and was ready to join the army and fight when his chief prospector returned from the field with a sample containing conspicuous ilmenite. It was late evening. They washed the sample themselves and could hardly believe it when a perfect 2-carat octahedral diamond crowned the heap of sieved concentrate. Early the next morning they returned to the geological survey trench from which the sample had come. They had soon recovered more diamonds.

A summer afternoon at Williamson.

There was no time to waste. Williamson drove as fast as he dared on the rough gravel track to the district commissioner's office in Shinyanga where he took out an exclusive prospecting licence. If he had been elated when he saw the diamond that evening, his credulity was severely tested when he set out to delineate the pipe. It was enormous, bigger than any he had read of, and certainly the diamond content was more than sufficient to start a mine.

News of the find soon filtered through to the War Office in Whitehall and Williamson found himself better supported by the various authorities he had to deal with in Tanganyika than he could ever have hoped for. To be entirely dependent for a drastically growing need for industrial diamonds on De Beers, as they had been, had concerned War Office strategists deeply. Now they had a supplementary supply. To get it from the drawing board to production became a priority.

In view of the rich concentration of diamonds at the surface, Williamson was able to fund building and expansion of the mine without having to use outside finance. The biggest kimberlite mine in the world was his and his alone, except for the shareholding he chose to allocate to his brother in Canada and his lawyer in Mwanza.

Williamson was just 50 when he was diagnosed with advanced throat cancer. He had never married and had lived a hard life, drinking quite heavily and keeping himself under constant pressure. His mine was everything to him, and the same dedication that saw him succeed at university like no one before or after him drove him to the end. He fought the cancer for a year but his number was up. He died on 8 January 1957.

The mine, still known alternatively as Mwadui, after the area where it is located, and Williamson, continues to operate, though the grade, now that the high-grade crater facies has been mined out, is not nearly as high as it was. After being owned by a De Beers–Tanzania government joint venture, it was fully nationalised by Tanzania's first president, Julius Nyerere, before passing back to De Beers. When De Beers failed to redeem the economics of the mine sufficiently to justify retaining it in the group, in 2009 it disposed of the mine to a junior South African company.

Lesotho

As John Williamson's first diamond was found in the shadow of a baobab tree, it is appropriate that he chose this iconic symbol of Africa for his mine's logo. Recently Lesotho's Letšeng-la-Terai diamond mine narrowly escaped acquiring the same tree for its emblem, despite the fact that the nearest baobab is many hundreds of kilometres away, in a very different part of Africa. In the end the mine's new majority shareholder, Gem Diamonds, whose simple logo depicts this tree, was persuaded to leave the mine's original symbol – a diamond crystal – in the Letšeng insignia.

To: Pappa From: kayla 6 April 2011

The two pipes of the Letšeng Mine, with the slimes dam beyond them.

There are, after all, similarities between the two mines. Each is unique in itself, Mwadui for the giant size of its pipe, Letšeng for the giant size of its diamonds, not to mention their exceptional quality. And more than any other mines in the world each is synonymous with one man, Mwadui with Dr John Williamson, Letšeng with Keith Whitelock.

Whitelock's first acquaintance with the kimberlite pipes of Lesotho took place in 1959, when he was sent by De Beers to be part of the team following up on the first phase of reconnaissance exploration for diamonds in the Maluti highlands – the Roof of Africa – in the pre-independence Mountain Kingdom that was Basutoland Protectorate. In a countrywide survey, the Basutoland-born mining entrepreneur Colonel Jack Scott had identified a large number of kimberlite pipes that needed to be pursued. Scott's investigation was triggered by the discovery by police looking into an accidental death that there were at least a hundred people digging for diamonds with some success in the Kao valley in the highlands. In 1955, the year after the police discovery, Scott was given permission to prospect the entire country. By 1959 four years of high-cost exploration had consumed as much of Scott's resources as he cared to devote to the project with no sign of a major diamond deposit to show for it. He invited De Beers to take the project further as joint venture partner.

Before he could start exploration in earnest, Scott had to have a base in the mountains where he expected to find the pipes he was looking for. Kao, with its known reserves of diamonds and by all indications a very large pipe (nearly 20 hectares) itself, was a good place to start. His first task was to get a track into Kao, so as to be able to transport his bulk-testing equipment into the area, both to start evaluating Kao and to assess any new kimberlites which his team discovered. That Scott was able to build a road into Kao at all, even if it did take two years, is testimony to his determination and resourcefulness. It is the most extremely rugged terrain in southern Africa, and its winter climate is harsh, with snow falling heavily more winters than not, and night-time temperatures plummeting to –10°C and below.

From the base-camp at Kao, Scott's geologists trekked through the precipitous wilderness, collecting stream sediment samples and following up reports of sands rich in the glittering black ilmenite, or *sekama*, used by the BaSotho people for personal ornamentation during ceremonial events. At an early stage it was established that any significant concentrations of ilmenite were derived from kimberlites. It was this more often than the garnets commonly used as indicator minerals elsewhere that quickened Scott's geologist's pulses. Yet we are told that it was garnets that the geologist Peter Nixon found in the sample he took from the swamp- or sponge-like depression half-way from Kao to the northeast-facing Drakensberg rampart that forms the border with South Africa. The local shepherds told him the place was known as Letšeng-la-Terai, the 'turn by the swamp', where the track they used for stock-trekking made a marked change of direction. Because the kimberlite is softer than the basalt which makes up the Maluti highlands, and into which it was intruded, it tends to weather

The sun may not bring much warmth to this scene.

KEITH WHITELOCK

You can take Keith Whitelock out of the mountains but you can't take the mountains out of him. In 1959 Keith started his career in diamonds, doing regional exploration in the towering Roof of Africa of Basutoland (now Lesotho), and that's where he is today, aged 77. Fifty years ago there were few roads scratching at the edge of the mountains, for the rest only pony trails. Now there are tar roads and ski slopes, and – the key to it all – the world's lowest-grade diamond mine, Letšeng-la-Terai. Keith was in charge of the mine from its opening in 1973 until De Beers decided to walk away ten years later. After managing the Orapa (Botswana) and CDM/Namdeb (Namibia) mines with great success, he found that retirement held no allure. He would get Letšeng started again. With the mine now running as an unqualified success, he has taken on a new developing mine not far away, at Kao. It's tough but Keith is no stranger to tough. Watch this space.

preferentially. It makes small basins which collect water, accentuating the weathering, and forming clearly recognisable 'sponges'. Nixon would not have been surprised to see kimberlite indicator minerals in his sample.

Follow-up of the Letšeng-la-Terai kimberlite – in fact a main pipe and a satellite – was disappointing. Thorough testing showed the grade was extremely low, a good deal lower than at Kao, for example. This was the state of the kimberlite investigation when De Beers and Keith Whitelock entered the fray. Keith was one of four geologists who led teams into all parts of Basutoland, sampling every stream for signs of kimberlite indicator minerals. Where these were found they were traced back to the pipes and fissures that were their source. Just under 60 such sources were found, which, in a country of around 30 000 square kilometres, gives it more such bodies per unit area than any other. But it was found that quantity and quality did not go hand in hand, and testing gave one disappointment after another.

Letšeng appeared to be no exception. Its low grade and remote location consigned it to the 'out' tray and in 1960 the Scott–De Beers joint venture agreed it could be opened for public digging. Word was put out that claims of three metres square could be worked on payment of 50 cents per month, and a mini-rush was soon under way. After a few top-colour stones of over 50 carats were recovered, the BaSotho diggers were joined by the more adventurous from South Africa. A hand-made air strip had been constructed at an early stage and in 1963 the government had a road constructed that allowed access in days rather than weeks.

In 1965 the amazed attention of the diamond world was drawn to Letšeng when a 527-carat diamond was discovered there. The focus sharpened two years later when a stone of 601 carats – the Lesotho Brown – was recovered, at the time the eleventh biggest diamond ever found. But the going was getting harder. Patches of soft surface material both on and downslope of the pipes were now few and far between, and crushing the unweathered kimberlite was not a proposition for diggers who mostly lived from hand to mouth.

The Basutoland Protectorate became independent Lesotho in 1966 and the new administration, aware of the parlous state of its fledgling economy, was anxious to see the deposit turned to account. Rio Tinto Zinc, or RTZ as it was known, had shown itself a world leader in turning low-grade deposits to account with the massive new copper mine at Phalaborwa in South Africa, and was embarking on a feasibility study of a huge low-grade uranium deposit in South West Africa that had been spurned by at least one South African mining house. Never one to walk away from a challenge, RTZ accepted the Lesotho government's invitation to investigate opening a full-scale mine at Letšeng.

RTZ bulk-sampled the main pipe for four years, from 1969 to 1972, both by underground 'drives' from the side of the hill and in open trenches, before deciding that the deposit did not fit their profile as a miner that could make low-grade deposits pay by optimising economies of scale. They had shown that Letšeng was low-grade – it was just too small, as well as being extremely remote and in the most demanding terrain.

A collection of premier Letšeng diamonds.

Would the RTZ–Letšeng marriage have endured if the miner had found a 500-carat, 'D'-colour flawless diamond in the concentrate one day and sold it for top dollars? We don't know. What we do know is that within a decade RTZ's Australian subsidiary, CRA (Conzinc Rio Tinto Australia), would enter the diamond business in a big way. At Letšeng, it didn't have to wait long to strike it lucky.

The next year, 1973, Anglo American, as prospector for De Beers, was back in Lesotho, invited by a government determined that the jewel in his humble crown should be recognised for what it was or might become. Keith Whitelock was in charge of the sampling team. In March 1975 De Beers announced a new mine would be added to its portfolio and development of the Letšeng-la-Terai mine started. By 1977 the mine was in production. But only for seven years. In 1983, prompted by an alarming downturn in diamond prices, the company was forced to review and rationalise its overall strategy. The marginal (at best) Lesotho mine was one of the first casualties. Once again Letšeng found itself in limbo, and Whitelock moved to the new De Beers mine on the plains of the Kalahari, Orapa, where he took over as general manager.

Twelve years later, now retired from De Beers and a free agent, he was back in the mountains. He opened up his house in the eastern Free State and set about applying for a mining licence for the Letšeng deposit for the company he formed, Letšeng Diamonds. It took five years, but in 1999 the rights were issued and the financing of the mine could proceed. In the end the company comprised a 24 per cent Lesotho government shareholding and 76 per cent industry investors, with Keith Whitelock as managing director.

While the construction of the open-pit kimberlite mine and its plant was under way, Whitelock brought in Alluvial Ventures, owned by old associates of his, the Cornelissen family. Father 'An', an old hand in alluvial diamond recovery, and his three children, Lourens, Elizabeth and Courtney, wasted no time. By late 2001 they had a compact and highly efficient operation in production to process gravels downslope from the kimberlite pipes. The first parcel of diamonds was sold early in 2002. In November that year, working a gravel that neither Whitelock nor An Cornelissen had felt had much promise, some 50 to 80 metres downstream from the main pipe, they recovered in a matter of days three 'D'-colour flawless stones of 214, 125 and 98 carats. They were sold for $5.5 million: today the price would be nearly treble.

They hardly dared hope it was a sign of things to come. They would soon know. By 2003 both the main and satellite pipes were in production. The initial mining investor disposed of its shareholding in 2006 to Gem Diamonds, with 6 per cent of the holding being transferred to the Lesotho government, so that Gem currently owns 70 per cent of the mine and the government 30 per cent. Whitelock was retained as CEO.

He and his team were not betrayed by that early portent. In August 2006, with mining

Run-of-mine Letšeng diamonds.

and processing fully under way, the first monster stone, a 603 carat, was recovered, which was duly named the Lesotho Promise. The Letšeng Legacy made world headlines a year later at 493 carats, and in September 2008 the 478-carat Light of Letšeng was sold for a record price for a white stone, of $38 500 per carat. Earlier that year a 26-carat pink diamond had realised $99 000 per carat. How had Whitelock and his team known? They had seen size frequency distribution (SFD) curves of Letšeng's run-of-mine production (a recent innovation in the evaluation of diamond deposits), which showed that big stones were to be expected, and that before the mine was worked out a 1000-carat diamond should be discovered. It is the only kimberlite mine in the world where the average stone size is a whopping 1.8 carats.

The point needs to be made that there are mines which knowingly crush large gems, accepting that risk as part of the trade-off between value of product and cost of processing. On the other hand, because of the very high value of Letšeng's diamonds, nearly twenty times higher than the world average, Whitelock saw the need to be sure that large stones remained intact, even if it required an extra circuit in his plant. It raised the cost, but he judged it a necessary addition, and he has been resoundingly vindicated.

Letšeng would be the jewel in any country's crown. Lesotho's coronet is, all the same, a humble affair. Quite close to Letšeng-la-Terai, the 20-hectare Kao pipe continues to tantalise investors with its marginal grade, which, though higher than Letšeng's, does not include any of the mega-diamonds that have become the hallmark of Letšeng. And there are a handful of other pipes which at the time of writing were in the stage of pre-feasibility evaluation. None is large, however, and although the production may include a sprinkling of 'D' colour stones, none has delivered sensational material like Letšeng, nor do their SFDs suggest that they will. But if quality is put above quantity, Lesotho can be proud of its single singular mine and its 'diamonds in the sky'.

Geophysical exploration for kimberlite pipes in Botswana.

Botswana

I smiled when I first saw the aerial photograph of Orapa, taken in the 1960s. The photograph was one of the set that covered the whole country and could be ordered from the Botswana survey offices if one was doing mineral exploration. They were in the public domain.

Why I smiled was that the pipe stands out so conspicuously. When I was employed in central Botswana as a young graduate in an extensive copper exploration project in the early 1970s, the mining company I worked for had a parallel operation looking for diamonds. Its *modus operandi* was to investigate and sample every 'circular feature' on a map compiled from a photo interpretation – from those same aerial photographs – of the concessions the company had been granted. The map had been made by a well-known photo-geologist from the United States and was a superb piece of work.

Our 'diamond team' didn't find any pipes, because there aren't any. Though we didn't appreciate the significance of it at the time – and I'm not sure that anyone would have at that time – we were 'off-craton', not in elephant country. Moreover, Orapa fell far to the east of our concessions.

That is all by way of a digression, because Orapa was found by soil- and stream-sampling techniques tried and tested over decades by De Beers. There was some good geology involved, too, by a combination of De Beers exploration manager Gavin Lamont and Chris Jennings, then with the government Geological Survey.

Small diamonds and garnets had been found in a shallow drainage feature which had been followed 'upstream' to what looked like its end at a low rise in the very gently undulating eastern Kalahari. There was no sign of a kimberlite pipe which might have been the source of the minerals: the trail had gone cold. It was then realised that what appeared to be the end of the old stream course was not that, in fact. It went on, over the rise and down the other side. The sampling team picked up the trail on the far side of the rise and followed it. In 1967 it led them to Orapa.

What constituted the clever geology was that Lamont and Jennings realised that the rise, with streams appearing to drain away from it in both directions, had been formed after the

Jwaneng mine, producing a lot of the world's gem diamonds.

stream, itself a very ancient feature. Originally the stream, like any and every stream, had flowed down the slope, and that was the only way it had ever flowed, ever could flow. It was hardly a slope, though, but the sort of gradient that needs instruments to show which way it tilts. After the stream's course formed, the climate dried. It was no longer a stream, just a memory. Then, forces deep underground lifted the ground's surface along an elongate arc, cutting right across the stream course.

For a short while the geologists had been fooled, though not for long. Jennings had studied under one of South Africa's foremost geomorphologists and knew about the subtle but important arches of uplift, first proposed by South Africa's geologist extraordinaire, Alex du Toit. He had never known how useful geomorphology would be.

Like most kimberlite pipes, the main Orapa pipe was one of a cluster. If there was one cluster under the vast Kalahari – although Orapa is right at the edge of the sand-filled basin – were there more? The search was intensified. Amongst other innovations, sampling techniques were refined to include the sampling of termite nests. Over many generations termites dig to remarkable depths, and some of the material they unearth finds its way to the surface. That is how the Jwaneng pipe, the world's richest deposit, was found, under 40 metres of calcrete and sand. Thanks to the Trojan work of nature's smallest underground miners, tiny grains of garnet and ilmenite were picked up by the De Beers samplers.

If the Letšeng mine in the bleak and icy heights of Lesotho stands proud for the value of its stones, Jwaneng, in the acacia-studded Kalahari savannah of southern Botswana, wins first prize for the value of the resource. Not only is it big – 54 hectares – but, at a recoverable grade of 141 carats per hundred tonnes, it is exceptionally rich. It is mainly Jwaneng and the 106-hectare Orapa that in the space of a few years have rocketed Botswana into top position (by value) of the world's diamond-producing countries.

As they say, the early bird catches the worm. None of the numerous kimberlite pipes found – mostly using sophisticated geophysical techniques – in Botswana since Jwaneng have been turned to account. One of the Orapa cluster of pipes that De Beers discovered and let go is having a mine built on it by a London-listed junior company, with De Beers participation. Others are slated for development in the (probably distant) future, where deep sand or less than spectacular grades require substantial diamond price improvement for the resources to be viable.

The bird that was De Beers had waited six years before Orapa. They might have been futile years like the many De Beers spent in Brazil. But they were not: the company was rewarded with worms plump beyond imagining.

The Udachnaya Mine on the Arctic Circle.

CHAPTER EIGHT

RUSSIA

In the encyclopaedic *International Diamond Annual* of 1971, the Soviet Union was listed as the world's leading producer of gem diamonds for 1970 – at an estimated 2.7 million carats, edging just ahead of South Africa. Yet had the publication appeared 20 years earlier, the Soviet Union would not have featured. Just past the half-way mark of the twentieth century a new diamond producer exploded onto the world stage.

In 1829, nearly 40 years before the Eureka diamond was discovered in South Africa, Pavel Popov found a diamond in his gold pan while he prospected for gold in the Urals. It was to be a false start, though. On the assumption that the stone had not travelled any great distance, a search for the mother lode combed the mountains. No source was found. Geologists argued the source must be an unusual rock formation called harzburgite, well known in the mountain chain, and genetically not far removed from the mantle rocks where diamonds originate. So the typical heavy minerals from these rocks were used as indicators in the sampling programme, and all harzburgite found was tested for diamonds. Sample after sample proved sterile.

Meanwhile, the discovery of the mother lode – which would come to be known as kimberlite – on the dry plains of South Africa identified for the first time the true family of the diamond, which was in fact quite different from what the Russian prospectors had been looking for. Even before the kimberlite was discovered, the South African diggers recognised the bright red rubies, as they called them – in fact, pyrope garnets – as a sure-fire sign that they were on the trail of diamonds. We know now that the reason for this is that pyrope and diamond have specific gravities so close that the two will tend to concentrate in the river environment in the same places. And garnet is at least an order of magnitude more common in most kimberlites than the gem that it shadows in its journey from deep in the mantle; hence its usefulness as an indicator.

It took the discovery of kimberlite ore in South Africa to prove to geologists and miners what they had always in their heart of hearts suspected. The diamonds found in river gravels, whether ancient or modern, did, after all, come from a primary source that could make great mechanised mines, not just the

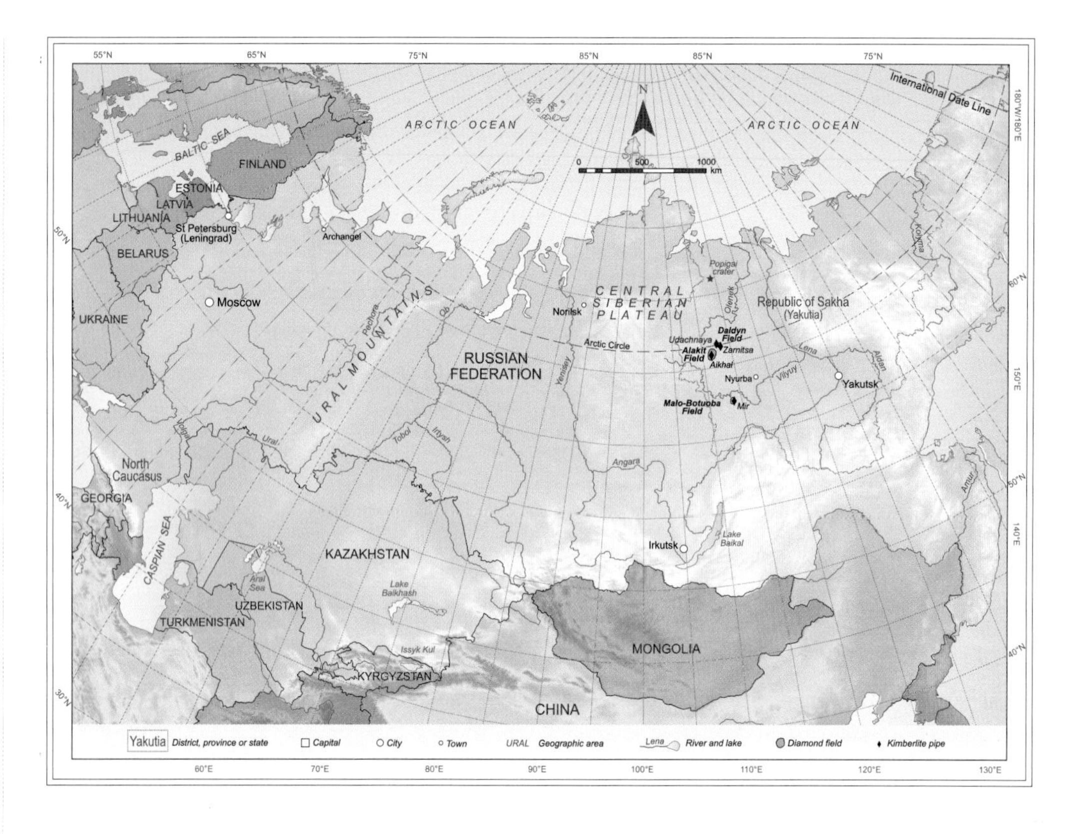

scratchings on the surface that the gravel diggings amounted to. And as industry burgeoned across Europe and North America, the diamonds needed for cutting and grinding, which kimberlite pipes could yield, gave the pipes extra lustre as a priority target in mining circles.

The Soviet Union was no exception. The great Communist dream needed them, the appetisers from the Urals said they were there somewhere, and the sheer size of the Soviet territory made it unthinkable that they couldn't be found. There was every reason to search for them. Motive was there in abundance: the *modus* was not so clear.

The early narrative of Soviet diamonds under Communism is one of false starts, of hands put up and shunned, and red herrings. In a couple of decades after the 1917 October Revolution, a gaping chasm had opened between Communist ideology and Western pragmatism. When the search for diamonds first appeared on the strategic agenda as a priority item, the way in which capitalist, colonial South Africa had found – and continued to find – its mega-deposits was irrelevant: the Soviets would do it their way.

This rejection of by now tried and tested technology was only part of the reason for the late entry of the Soviet Union into the global diamond arena. The proliferation of bureaucracy under Communism was quite as big a factor. Whichever historical summary of the Soviet diamond industry you read – and there are several – what strikes one is that it was a highly politicised business. Councils, commissariats, commissions all flourished, each with its own agenda.

Undoubtedly, the greatest brake that slowed progress was the terrain itself. Siberia, where the search would be mounted, was about as inhospitable a patch as Earth has to offer. Then devoid of infrastructure and access because of its remoteness, and – now as then – so bitterly cold for most of the year that it challenges life itself, it called for monumental effort and determination. At these latitudes snow and ice are everywhere for seven months of the year. Everyday materials take on an entirely new character. Iron bars snap like matchsticks, rubber breaks like crockery and frozen bread has to be chopped by axe.

As early as 1929 the Soviet geological fraternity was on the verge of making the vital connection between pyrope garnets – the kimberlite prospector's best friend – and diamonds. It was a clue that, if embraced, would have put them firmly on the trail. But it was not to be. We are told that suggestions of similarity between South Africa and the Soviet Union were rejected when two 'well-known' geologists who had studied kimberlites in South Africa that year were arrested after their return.

In 1938 the findings of a Russian delegate to the International Geological Conference in South Africa were summarised thus in an official report: 'Pyrope is the key to finding diamond deposits.' Too important to be left lying around, the report was classified as Top Secret and locked away in a safe. There it lay. The Soviet trail had gone cold.

At about the same time G.G. Moor, while on the Arctic Expedition which traversed northern Siberia, found a kimberlite-like rock we now call lamprophyre. Astute enough to assume the kinship of the two rock-types, Moor wondered if this might be telling him that he had crossed diamond country. When Vladimir S. Sobolev – one of the giants of Soviet geology – read Moor's report in 1941 he was bold enough to say, 'The Central Siberian geological platform is to all intents and purposes identical with the South African diamond-bearing region. Here if anywhere, we will find diamonds. I particularly direct attention to the valley of the Vilyuy River and recommend that detailed prospecting be started there at once.'

What wizardry led him to such a conclusion? Sobolev's early *magnum opus* had been a 214-page paper published in 1936 on the 'traps' or basaltic lavas – also called flood basalts – that occupy a large part of the Siberian Plateau. The Siberian trap province is the largest of the vast fields of flood basalts known around the world. Others are known in India, southern Africa and southern Brazil. Sobolev would have studied all the literature he could find relating to those. In those days, particularly in the world of academia, there was a free, albeit unilateral, flow of information between the West and the Communist world, so he would have had a clear picture in his mind's eye of the geology of those regions.

Sobolev recognised enough similarity between the stable 'shield' areas of diamond-producing South Africa and his own Siberia, both essentially undisturbed for three billion

Siberia, beautiful on a clear day but bitterly cold through the long winter.

years, to rank the shield area there number one as a diamond exploration target. Two and a half decades before Tom Clifford's classic 1966 paper proposing that diamond-bearing kimberlites are found only on stable 'cratons', Sobolev had intuitively understood the same linkage, and applied it to his home turf.

Sobolev's discovery was made in the early years of the Second World War, when the Soviet economy was geared entirely towards the large-scale production of armaments, heavily dependent on industrial diamonds. After 1946, with the Cold War and the isolation it brought just beginning, a domestic supply of diamonds had become a priority of the greatest urgency.

Let us look at what diamonds are used for in industry. In 1946 diamond-impregnated tools were used 'to "true" the precision tools that grind gun bores and airplane and automobile crankshafts, valves and gears; to saw and prove plastics and ebonite sheets for insulators and time switches; to grind pistons, valve seats and lenses for binoculars, range-finders, bomb sites and navigation instruments; to test the hardness of ball and roller bearings and rings; to draw precision wire for electric transmission, minesweeping cables, airplane control cables and bracing wires, plane detectors, radio valve filaments, magneto windings and surgical uses; to help create hypodermic needles and ship hawsers; to help produce high-flying bombing fortresses, giant tanks that can crush other tanks, trees and fences and men, and crumple the unharvested corn'.[1]

Search operations for diamonds got under way in the Urals, Siberia, Kazakhstan and the North Caucasus. In the summer of 1949 a young field geologist, Gregory Feinstein, and his Amakinsky Expedition collected six diamonds from the Vilyuy River in Siberia, far to the east of the Urals where the early alluvial diamonds had been recovered. Was not this the area G.G. Moor had drawn attention to and the revered Vladimir Sobolev had recommended for

An exploration base camp in the winter of 1985.

follow-up? No one dared say it, but this was starting to look more like Kimberley – with its Vaal and Orange river diggings preceding the kimberlite discoveries – even if it was a sub-Arctic Kimberley.

More and more diamonds were found in river gravels in the Republic of Yakutia (now Sakha) in the central-eastern part of the country, but there was no pattern to the discoveries and for years the source remained elusive. The Politburo in Moscow ground their teeth: diamonds were being found but not the cornucopia they sought.

In the field season of 1953 three pivotal breakthroughs were made. Firstly Natalia Sarsadskikh, chief of one of the field parties, identified an area far to the north of the main Vilyuy headwaters as structurally favourable for the emplacement of kimberlites. Then came the realisation that the ruby-coloured grains that her assistant Larissa Popugaeva had found, together with a small diamond, in one of her panned samples might be critically important. Neither had seen this mineral in concentrates before. In Leningrad Natalia's husband and mentor at the local university, Alexander Alexandrovich Kukharenko, confirmed that the grains were indeed pyrope garnet. Lastly the two women realised that the garnets were kimberlite indicators and that they must have come from streams draining kimberlite. Any stream with pyrope garnets should be followed until there were no more indicators, at which point they would know that they were very close to their kimberlitic source.

When the time came to return to the field, Natalia remained behind in Leningrad, where her daughter was born in February 1954. Appalled at the prospect of going into the field on her own to follow up their discovery of the previous summer, Larissa tried to cry off, for reasons that can be gleaned from her diary of that first field season. 'But already the workers are grumbling, I am fed up with them, they are worse than bitter horseradish. They are clumsy, evil and lazy, especially M.A., who brags unmercifully, a know-all without any understanding. It is nauseating … We cook porridge, jelly, we make flat bread, since the ordinary bread is finished. In the daytime frequently I think about home. How are my good folk? How is my dearest, beloved daughter, is she ok? I want to be home, the quicker the better. The gnat is biting … Brrr!'

Some days later: 'Rapids appeared, then the river turned almost dry, but the boat is heavily loaded. But we had no alternative but to drag the boat over the rapids and dry places because there was no help. We had to drag until the evening. There was the danger of damaging the boat on the sharp stones of trap rocks. The water is freezing cold, but we have to go into the middle of the river up to our knees. My legs were deadly cold. Thus we dragged until dark. On the banks there were many tracks of moose, but we are still fasting.'

It was with memories like this that she set out the following spring with her trusty field assistant, Fedor A. Belikov, and his dog. The small party was landed by a hardy little biplane on a sand bar in the river where the garnets were found. They were completely self-sufficient, not a soul for hundreds of kilometres in any direction. They worked upstream, sampling the sediment every 500 metres, pulling their gear in a rubber dinghy which they tethered with a long rope to a tree up the river while they stopped and searched. Like this they covered two kilometres a day, for each sample digging 300 small bucket-loads of sediment out of the icy river, then washing it to concentrate the heavy minerals. They were on the Arctic Circle and, though the summer days were warm, as the sun got low the air chilled to near zero.

They found lots of garnets in every sample until suddenly there were practically none. They went back to the small tributary stream between the last sample and the present one and followed it up. Sediment was sparse but it didn't matter: even a few handfuls gave them more than enough garnets. The stream splayed out into a swamp. Above the swamp no more garnets. They turned back and sampled the soil on the sides of the stream.

On 21 August 1954 they worked on a low flattish rise, sparsely wooded with larch and alder trees, less than two kilometres from the stream. At about midday it started to rain. Leaving Fedor to build a fire, Larissa sat, wondering whether they were close, her eyes unfocused on the ground in front of her. Suddenly, almost involuntarily, they focused. She shouted, 'Fedyunya, look! Blue earth, and entirely full of pyropes.' The Soviet Union's first kimberlite pipe had been found.

The Zarnitsa (variously translated as Summer Lightning, Dawn and Thunderflash) pipe, as it came to be known, though quite large (21.5 hectares), was not richly diamondiferous. It was eclipsed the following year first by the Mir (Peace) pipe, 240 kilometres southwest of Zarnitsa, and days later by the Udachnaya (Lucky) pipe, in the same kimberlite field as Zarnitsa.

In 1955 the geologist Yuri Khabardin was a member of the team following up the report by a young student, Vladimir Shchukin, of a pipe far to the south of Zarnitsa. Imagine him

Vladimir Shchukin, now doyen of Russian diamond geology, was in the 1950s a young field geologist fired with the excitement of what he was finding.

VLADIMIR SHCHUKIN

At 80, Vladimir Shchukin describes himself as being full of life, strong and looking forward with optimism. He still consults for various diamond exploration companies and his earlier master, Alrosa, still calls on the patriarch's expertise and wisdom regularly. In the early 1950s he joined the Amakinsky Expedition in its diamond search in southern Yakutia and was commissioned in 1955 to investigate the recently discovered Zarnitsa kimberlite pipe, Russia's first. Later, exploring what would prove to be the most fertile part of the Siberian shield for diamonds, Shchukin and his team located the Udachnaya pipe, one of the big high-value pipes in the world, and a series of other important ones. In 1969 his team found the International pipe, by far the highest-grade pipe known anywhere. He went on to take an increasingly active role in strategic planning for the state miner, Alrosa. In recognition of his contribution he was awarded the honorary title of Lenin Prize laureate, the Order of the Red Banner of Labour, the Order of the Polar Star (Yakutia) and the Order of Lenin medal, the highest honour bestowed by the Soviet Union. Russia is proud of him.

fighting to contain his excitement as he radioed in to his office to give them the code message, 'I am smoking the pipe of peace.' Evidently this could be taken by his superiors to mean that Khabardin had found, and positively tested, a kimberlite pipe. It was a historic radio call.

Foxes feature prominently in Russian diamond legend. Yuri Khabardin is reported to have found the first trace of the Mir kimberlite at the mouth of a fox-hole. Another legend has it that Larissa found Zarnitsa after shooting a fox with blue earth on its belly, which she was sure must come from kimberlite, but the story appears apocryphal. Foxes aside, the two discoveries were each supremely important in their own way. Zarnitsa was to point the way to the main Siberian diamond region, the adjoining Daldyn and Alakit fields, and Mir, in the more southerly Malo-Botuoba field, would be the main Russian producer for decades.

Interestingly, the Mir pipe was the first kimberlite to be mined outside Africa, other than very minor production from Indian pipes. This congruence between Africa and Russia calls to mind the early cynics who scoffed at the notion of diamonds being found in the Cape because the country 'looked' wrong. How they would have shaken their heads at two places as dissimilar as the Siberian taiga and the blistering plains of the Kalahari producing diamonds from rocks that were to all intents and purposes identical.

Larissa's lament

Despite her historic discovery, Larissa Popugaeva was to sink slowly into obscurity, collapsing on a street corner with a massive aneurism of the aorta, lonely and dispirited, on 19 September 1977. A 29.4-carat diamond was named after her, streets in Udachnaya and Aikhal bear her name, and a monument honouring her has been erected in Udachnaya. I hope she smiled as she heard – long after her discovery and thanks only to an insistent reporter's recommendation – that she was to be awarded the Order of Lenin for her achievement.

Larissa's case may have been extreme or it may have been typical: we don't know. What we do know is that instead of the 'flowers, fanfares, ovations' she might have expected when she returned to the headquarters of field operations in Nyurba, having outlined the pipe by pitting, her reception was cool. Confusion and professional jealousy seethed just below the surface. In the next field season, using previous field data, she pointed Vladimir Shchukin to the Udachnaya pipe, which he found within days, and shortly after Khabardin had discovered Mir. For her trouble she was discharged from the Amakinsky Expedition.

◆◆◆◆◆

BY 1955, as we have seen, the Soviet Union had staked a major claim to fame in the world of diamonds. Already in 1959, it made its first appearance on the international diamond market, presenting 13 000 carats of rough gems for sale. The next year a third major kimberlite pipe was found at Aikhal, not far from Udachnaya, where a payable pipe had been found in 1955, in what was becoming known in diamond-mining circles as the Daldyn-Alakit field. The Malo-Botuoba field, around Mir, was not as forthcoming as the more northerly field, the small 23rd Party Congress pipe coming into production in 1966 and the International pipe in 1971.

But by 1960 three large, high-grade kimberlite pipes had been found – Mir, Udachnaya and Aikhal – and the world was seeing the calibre of the stones. They were top quality. Those in the West who had hoped that the Soviet Union's progress to becoming a major world power, politically and economically, might be retarded by its dearth of diamonds had been confounded. And De Beers was wondering. The gems it had bought from the Soviet Union were good enough to be easily disposed of, and might continue to be so – or they would be if the market were regulated by normal forces. But with a top-end luxury item where the beauty is entirely in the eye of the beholder, an *embarras de richesse* had been an ever-present risk for nearly a hundred years. No analyst had the remotest hope of gauging whether the following year's output might be 130 000 carats and the year after that a million.

As the new fields opened up in Africa in the 1960s – in Tanganyika, the Congo and Angola – De Beers personnel were free to go and see for themselves. They could not do so with the Soviet Union. The rare straws of information grasped at told of fields deep in Siberia, far beyond the Urals, and as far behind the impenetrable curtain as it was possible

The Mir ('Peace') Mine, unusually close to its town, Mirny.

to be. Grainy black-and-white photographs from high-flying spy-planes showed the mines were small, smaller than Premier or the giant Williamson pipe at Mwadui in Tanganyika, not much bigger in fact than the Kimberley mines. As to what conditions were like so close to the Arctic Circle, the world could only speculate.

It was clear that, apart from the cold these sub-Arctic miners would have to contend with, there was the remoteness of the operation. The manager of the Mirny Diamond Administration, Viktor I. Tikhonov, writing in the 1960s, knew how that cost was built into each and every aspect of his mine: 'Everything from matches to excavating machinery was supplied over thousands of miles by way of the Trans-Siberian Railway, then hundreds of miles down the Lena River and, finally, through swamps and forests to Mirny. And many supplies are flown in, because the Lena River is ice-free for only four months a year.'

The same Mr Tikhonov was quoted as saying, in 1968, 'We call ourselves the country's foreign exchange department.' So, apart from the higher salaries than they would have earned anywhere else in the country, it was pride that drew tens of thousands of people to the wastes of northern Siberia: pride, ideological zeal and patriotism. They were, after all, waging a war not only against the cold; they were engaged in the war to end all wars, a Cold War.

The West had already found how hard that passion was to defend against. And in the world of diamonds De Beers was finding itself not so much in a war as in a poker game against the master deceiver, who smiled as though he had a royal flush, but could not possibly have – or could he? Realising that the game was altogether too dangerous, De Beers opened a dialogue with Moscow.

The man entrusted with negotiating a deal with the Soviets was Philip Oppenheimer of the CSO, cousin of the De Beers chairman, Harry. De Beers would buy 100 per cent of the gem-quality rough from the Siberian mines at a premium, he proposed. As the advantage to everyone in the multi-linked diamond chain, from miner to wearer, of single-channel marketing was so manifest, the principle was hard to resist, whatever one's political persuasion. Nevertheless, the Soviet Union could not be seen to be trading with the West, particularly with an organisation widely known to be intimately connected with racist South Africa. To consummate the deal, corporate camouflage from Siberia to London would have to be put in place – companies within companies with connecting linkages that were long and devious. In the end the cover was impenetrable. Moscow could continue to denounce the monopoly in apartheid South Africa at the same time that its gems were headed for 17 Charterhouse Street – the very symbol of capitalism.

The Diamond Trading Company's Sir Philip Oppenheimer was largely instrumental in arranging the sale of early Russian diamonds through the DTC.

De Beers, its Diamond Trading Company and Central Selling Organisation knew full well that, depending on how big the Soviet production was, they might have to stockpile rough, perhaps quite a lot of it. They had done it with their own stones before, and knew it was a price that had to be paid to keep the market stable. In the end it was well worth it.

The big imponderable was how large the production was likely to be. The aerial photos showed quite small pipes, which might, like many of the known pipes elsewhere in the world, be richer at the surface than deeper down. And as the mines went down and more waste rock had to be stripped to get at the kimberlite, normal economic forces would place a limit on how deep they could mine. If the De Beers strategists had seen huge open pits, like the Williamson mine, they might have been worried, but the modestly sized openings in the permafrost hardly seemed to pose a threat.

After ten years, the stream of stones reaching Charterhouse Street along its labyrinthine course showed no sign of abating. How could it be? Now we know the answer, even if it's not the whole truth. The Siberian pipes are high-grade, some of them phenomenally so. In making assumptions as to the longevity or sustainability of the Siberian mines, it had been reckoned that the grades would be similar to the South African pipes, which are mined if they run to more than 15 carats per hundred tonnes (cpht), though some of them are twice as rich as that, and one or two even more. The International pipe, near Mir, is reported[2] to have worked kimberlite with a grade of 10 carats per tonne (i.e. 1000 cpht) from 1969 to 1980, and even the early mines of Mir and Udachnaya reportedly recovered 200 and 120 cpht respectively.

Though by far the majority of Russian production has been from kimberlite pipes, the scree deposits downslope from the pipes have proved a prolific source of diamonds, as have the river gravels, further from the pipes but connected to them nonetheless. Because they lie at the surface, these unconsolidated accumulations can be economically mined on a massive scale, and have proved an important supplement to the hard-rock mines.

An interesting footnote to the remarks about the grade of the Russian pipes is that the diamond plants recover every diamond in the pipe, no matter how small. This is not the case elsewhere because most mines outside Russia exist for their gems, which can be recovered, together with a lot of quite small industrial stones, by physical processes. The greater density of diamond relative to the material it occurs in, its tendency to fluoresce under X-ray, and its non-wettability, which allows it – and it alone – to adhere to grease when sprayed with water, all combine to make the diamond recoverable in a chemical-free circuit. This offers an enormous cost advantage.

In general the Russian diamonds are not big. Over the years the biggest stones recovered were the 239.05-carat Star of Yakutia in 1973, the 342-carat 26th Congress of the Communist Party of the Soviet Union in 1980, and the 242-carat Free Russia in 1991. In the global context, and considering the scale of Soviet production, that is not a lot.

On the other hand, Siberian kimberlites produce a far higher proportion of perfect octahedra with knife-sharp edges and needle-like points than, for example, South African kimberlites and their alluvials. The Soviet rough is more uniformly sized, too, than run-of-mine diamonds in Africa, being mostly between 1 and 4 millimetres in diameter, though of course rare bigger stones do turn up and smaller than 1 millimetre are certainly not unusual. A consequence of the size uniformity is that the Russian beneficiation industry – cutting and polishing – is able to offer a very standard product to the jewellery trade. These small stones are called Silver Bears, and there is every indication that even before cutting begins, the rough is carefully sorted for size and quality, aiming to fill a niche in the market that looks for high levels of consistency.

The perfect octahedron in the centre of this group is typical of a major part of Russian production.

We have seen how the need to engage with each other was clear to both Moscow and De Beers from the outset. But that was over 40 years ago, in another millennium. The question everyone asked was how the relationship so discreetly established as the Soviet Union entered the fray in 1959 would change as they grew their operations and as – perhaps, who knew? – De Beers drew towards the end of its own resources and its global influence in the marketplace waned. Would the balance tip and, if it did, what would happen to the relationship?

During all those decades, as *glasnost* and *perestroika* became household words, as De Beers found massive new diamond deposits in Botswana and as newly decolonised African countries used their independence, the lines of communications were kept open. Heated words were often exchanged, but pragmatism prevailed. The Soviet Union, then the Russian Federation, continued to sell a great part of its production to the Diamond Trading Company.

In the meantime, though, the Russian parastatal diamond company, Alrosa (an acronym for Almazy [diamond] Rossii-Sakha [Russia-Sakha]), sold both rough and polished goods in Antwerp and elsewhere on a growing scale, always careful not to destabilise prices by oversupplying the market. De Beers and the Diamond Trading Company were changing their strategy, too, in recognition of what was happening not only in Russia but in Africa, Canada and Australia as well.

Alrosa and De Beers were not dealing in a vacuum, though, and forces beyond the control of either of them would be brought to bear. Consider, for example, this headline on the Internet website CommodityOnline on 12 May 2009, drawn from the *New York Times*: 'Forget De Beers, Russia to decide diamond prices.' The key to the point about outside influences occurred well down in the article: 'The recession also coincided with a settlement with European Union antitrust authorities that ended a longtime De Beers policy of stockpiling diamonds, in cooperation with Alrosa, to keep prices up.'

The long and the short of it is that while De Beers, a commercial company like any other, puts some of its major mines on care-and-maintenance to cut its operating costs and avoid a glut of unmarketable rough, Alrosa is forced to keep mines open by a Moscow 'gravely concerned about potential unrest by disgruntled unemployed workers'. Alrosa is the main employer in the Republic of Sakha, and it provides work for significantly more people than De Beers. For the first time in history Alrosa will be producing more diamonds than De Beers.

The Alrosa name proudly written in diamonds.

What underlies the Internet headline is no fundamental disharmony between the two companies. It is just that their drastically different corporate profiles require strongly contrasting strategies to deal with the temporary problem of potential oversupply. The problem is acutely exacerbated by the European Union's meddling in a century-old tactic: stockpiling rough to regulate its price. Since the first time it was used, it has consistently worked to the advantage of every aspect of the industry and consumers alike. The diamond industry will survive the division.

In the long term the Russian diamond business is in good shape. A new field is being opened in northwestern Russia around Archangel and medium-grade deposits in the established fields in Siberia await the inevitable long-term improvement in prices of rough before they are turned to account. We will continue to see a stream of top-quality diamonds coming from that vast country with its several cratons for decades to come.

If we should see an anniversary edition of the *International Diamond Annual* in 2021, we shouldn't be surprised to discover the Russian Federation still topping the list of gem producers.

De Beers' Victor Mine in Ontario at an early stage of building.

CHAPTER NINE

NORTH AMERICA

Of the major diamond producers, Canada follows Russia, one at the beginning of the second half of the twentieth century, the other at the end. In both cases the events leading up to the discovery of richly diamondiferous kimberlite pipes read like fast-paced adventure thrillers. Other than that, and the fact that both searches took their operatives close to and beyond the Arctic Circle, they could hardly have been more different.

Whereas one was a monstrous state-backed operation, run like a military campaign, the other was waged entirely in the private sector, the only similarity with armed combat being the extent of subterfuge and espionage. In the early days of Soviet exploration the intrigue and rivalry were internecine; in Canada it was a frantic race between competing companies. Whereas the Siberian terrain was topographically rugged, with wide rivers carving valleys and carrying diamonds far from their sources, the Barren Lands of Canada were lake-studded flatlands, mostly covered with glacial till. Here the indicator minerals were transported by huge continental glaciers thousands of metres thick and then sometimes redistributed or even concentrated by great eskers or high-energy torrents of sand and boulders starting as lakes on the top of retreating glaciers and flowing from the top or emerging from crevices at the base of these huge ice sheets. Interpretation of the source of these indicator minerals was sometimes rendered very difficult as transfer of the minerals could have taken place between overlying till sheets originating from widely sundered elevated sources. And while Russian geologists could hope to see the diamond-bearing kimberlite outcropping, their counterparts over the Pole had to wait for rock-core to be drilled from below the scree or winter ice to know what their mother lode looked like.

◆◆◆◆◆

DIAMONDS HAVE BEEN showing up in the United States for over 170 years, from California to Wisconsin, not regularly but from time to time. In the early days, one should remember, there was no reason for North Americans to pay much attention to diamonds as India and then Brazil supplied much of the world's needs. Yet as the Brazilian industry grew and new diamond fields were discovered far and wide in that country, gradually an awareness took root in North America that they might be found just about anywhere.

To begin with, Brazilian rough did not reach New York directly, as there was no significant cutting facility there. Then, around the 1850s cutters started arriving from the Low Countries, and cutting and polishing of directly imported Brazilian and later South African rough began, first in Boston and then New York, where even today most of the world's largest diamonds are cut. By the end of the nineteenth century the United States had become a player in the young world diamond industry and North Americans began to take an interest in diamonds.

In the small collection of semi-precious stones I have collected over the years is quite a big, faceted, pink-violet kunzite from Minas Gerais. It is a rare gemstone and beautiful, though not nearly so valuable as even a much smaller diamond would be. The mineral – a lithium alumino-silicate – was named in 1903 to honour a celebrated American gemologist, George F. Kunz, and he is the reason I opportunistically refer to my personal treasury. George Frederic Kunz pursued diamonds in the United States with enormous zeal in the 1800s and early 1900s. Born in Hoboken, New Jersey, in 1856, he followed the fascination that small boys have with stones for his entire life and with singular passion. Among a list of distinctions and accomplishments too long to name, Kunz assembled a research collection of minerals for Thomas Edison as well as the famous Morgan-Tiffany Collection which ended up in the American Museum of Natural History. He was associated virtually all his adult life with Tiffany & Co., of which he became vice-president, and presided over the Kimberley Exposition in 1892. By the time he died in 1932 he had written some 500 books, papers and reports on mineralogy and had been awarded doctorates by several universities and the highest honours by the governments of France, Norway and Japan. George Kunz was truly a giant of American mineralogy and gemology.

He knew of sapphires in Montana and emeralds in North Carolina. He found pyrope garnets in Navajo anthills. Where were the diamonds? The mother lode for diamonds, and pyrope garnet, amply described soon after its 1871 discovery in the Cape Colony, was named in 1888. Kunz knew he first had to find kimberlite; then he might find the diamonds in the quantity his employers wanted. He saw diamonds from North Carolina, Montana, Alabama, Tennessee and Nebraska – but only singly and no kimberlite. Because of the Great Diamond Hoax that rocked the United States in 1872, one of the most brazen swindles of all time, which left a legacy of deep scepticism about any reputed diamond discovery, Kunz treated every report with the utmost caution. It took a number of discoveries in the northernmost states (from Wisconsin to Ohio) and, more importantly, the involvement of a geologist from the University of Wisconsin in Madison, William Herbert Hobbs, to convince him that they might be genuine and that they might be telling a story.

Together Hobbs and Kunz pondered the location of the finds and coupled them to the work of the Geological Survey of Canada whose geologists were studying the movements of ice sheets during the Last Ice Age. It was not difficult to show that the ice had travelled from Hudson Bay southwards, with two main lobes splitting to reach down into Ohio and Indiana on the east side of Lake Michigan and Wisconsin and Iowa on the west. At this stage, although diamonds had been found in the moraines of both lobes in the United States, none had been discovered in Canada. The discoveries in Wisconsin and east spawned another hoax in 1905, this time with Kunz's name wrongly attached to it. It was enough to take the wind out of the sails of Hobbs, Kunz and the Canadians, and thereafter diamond exploration around and north of the Great Lakes stalled.

Top: George F. Kunz admires a crystal of the mineral that bears his name. Bottom: One of the finest kunzite crystals known.

In the meantime American engineers were plying their trade in Kimberley, by then established as the diamond centre of the world. Some, like Gardner F. Williams, later general manager of De Beers Consolidated Mines, would go to stay. John Fuller spent long enough there to become manager of the Dutoitspan Mine before returning to the United States in 1908. He applied what he had learned in the Cape to America's first diamond mine, near the village of Murfreesboro, in western Arkansas close to the border with Oklahoma. Two years before, Kunz had confirmed two stones sent him from Murfreesboro as diamonds. By the time he and a geologist had reached the mine where the diamonds had come from, over a hundred more had been taken out of the weathered rock from the 30-hectare pipe and the

soil overlying it. After being shown a diamond *in situ* in the mother lode, he could no longer doubt the authenticity of the pipe and pronounced on its future in the most glowing terms.

Ownership of the pipe was complicated by the fact that a farm boundary ran through it. The Arkansas Diamond Company (ADC) was formed to take ownership of one side of the deposit. It produced diamonds on a small but significant scale, even if both mining and processing of the 'ore' used primitive methods. Next door no one was quite sure about the extent of the production. Until Fuller arrived from South Africa, the ADC's operation was chaotic. Then, with major systematic excavation in progress and an expensive washing plant constructed, the venture went from bad to worse. The gems were small, the bigger stones of industrial quality only, recovery was poor, and the money ran out. It was a disaster. In 1912 the ADC folded.

Kunz could not but be tainted by the mine's failure. Not only had he believed in it, but his belief had led him to invest quite substantially in it. As if his financial loss was not enough, in the year of the mine's closure his wife of 30 years died, and within a short space of time two of his three daughters had joined her. Lean, lonely years followed, but by 1920 George Kunz, soon to be remarried, was back in his stride even if, at 64, it was a slightly slower stride. A Canadian jeweller took him a 33-carat diamond – Canada's first – found while excavating a cutting for the railway between Toronto and Ottawa. Kunz's theory that diamonds of the northern United States had been transported by ice sheets from the north of Canada was on the way to being vindicated. What was more, he took his argument to its logical conclusion, predicting that 'a diamond mine or mines of great value are to be found in Canada'. George Kunz died in 1932, aged 76, leaving a giant-sized legacy behind him.

We turn back to Arkansas, to the Murfreesboro pipe with its division into two parts by a farm boundary fence. While the Arkansas Diamond Company lost its shareholders all their money proving their mine was not viable, on the other side of the fence events took a series of equally disastrous turns, only on a smaller scale. Howard Millar, who had worked his part of

The Crater of Diamonds Mine outside Murfreesboro in Arkansas is testimony to man's unquenchable hopefulness.

the pipe from 1912 to 1919 with exemplary determination and against the odds, returned 30 years later to try his luck again. This time he used a different approach: he named his prospect the Crater of Diamonds and opened it to the public. Millar's persistence was rewarded. The easy accessibility of the place, coupled with the discovery of occasional good diamonds and humanity's unquenchable optimism, made his venture a success from the day he reopened.

Soon after Millar's return, Murfreesboro took on a different relevance in the unfolding narrative of North American diamonds. As the United States' only diamond-bearing pipe, one open to all comers, it made an ideal reference point for anyone wanting to mount a systematic search for undiscovered fertile pipes. In the 1960s and 1970s the age of technology arrived in the industry, and diamond exploration geologists were broadening their horizon to take in geophysics and geochemistry. A known diamondiferous pipe was a good place to test their systems.

By now the concept of the multinational exploration company was gaining currency. Cash-flush oil giants like Shell, Mobil and BP were turning their hand to minerals, while previously localised and single-commodity mining companies gained a new, wider vision of the future. All minerals were targeted in greater or lesser measure, but for the first time it was realised that the monopoly De Beers had held in diamonds for nearly a century was not written in stone. Prices had firmed strongly and no new discoveries seemed to be coming out of South Africa, though it was rumoured that new pipes had been discovered in Botswana, not only by De Beers. If we were to put a date on the beginning of the great corporate hunt for diamonds in North America, it would be in the early to mid-1970s.

Diamond mining in the early days in Namibia used Sherman tanks, named after William Tecumseh, the older brother of John Sherman, whose Anti-Trust Act, passed in 1890, would outlaw entry into the US of De Beers operatives.

Suddenly the Crater of Diamonds had its fair share of visitors who were there not mainly in search of the shiny gems. Geologists wanted to see the North American diamond-bearing kimberlite for themselves. Those from De Beers were conspicuous by their absence, at least overtly, because of a piece of legislation called the Sherman Anti-Trust Act, passed in 1890 and named after John Sherman, chairman of the Senate Finance Committee. John's older brother was William Tecumseh Sherman, Union Army general during the American Civil War, after whom the celebrated Second World War Sherman tank was named. The Act's purpose and its effect were to outlaw the anti-competitive practice that is the backbone of cartels and monopolies.

De Beers vs the US

The monopolistic structure of the diamond industry had been a major thorn in the side of the United States during the Second World War. Diamond-tooled weaponry was essential to an Allied victory, but with De Beers and its Central Selling Organisation holding all the cards, procurement of the necessary diamonds was far from assured.

An early 1944 memo from the US Department of Justice to the Attorney General clearly illustrates the American ire. It states: 'Industrial diamonds are vital to the war effort,' and a little further on, 'The cartel controls 95% of the world's production … It refused to sell a large quantity of diamonds to United States Government agencies for a stockpile and refused to establish a stockpile of its own here.' The memo is headed: 'Proposed Equity Suit Against De Beers Consolidated Mines, Ltd., et al.'[1]

The stand-off was taken to another level 30 years later when the De Beers Industrial Diamonds Division was served with an indictment accusing it and two New York companies of price-fixing. De Beers did not plead, but agreed to desist from carrying on any further business in the United States. It is no surprise that the visitors' book at the Crater of Diamonds in the 1960s and 1970s does not show the signatures of any De Beers geologists. They knew where they were not welcome.

One of many North Americans to have worked for De Beers over the years was the Canadian Mousseau Tremblay. When Harry Oppenheimer bought the Williamson Mine in Tanzania shortly after its discoverer and owner's death, he persuaded Tremblay to stay on. He would report to the group consulting geologist for Anglo American and De Beers, Baltimore-born Dr Arnold E. Waters Jr. By 1960 it seemed that John Williamson had found the only payable kimberlite pipe in the Tanzanian field, and with his idol gone, Tremblay decided to return to Canada, joining the Geological Survey of Canada (GSC) in Ottawa.

The idea of 'going home' was also on the mind of Tremblay's ex-boss, except that for Waters in the early 1960s retirement was at the back of his mind. With the Russians rumoured to be embarking on a major diamond-mining project in Siberia, with Williamson having stolen a march on them in Tanzania and with British-based Selection Trust producing a steady stream of diamonds from their Sierra Leone mines, De Beers' stranglehold on diamonds had lost some of its grip. A major global expansion was called for. Arnold E. Waters Jr. was at the forefront of its planning as Mosseau Tremblay settled into his GSC office in Ottawa.

Waters and his team, in sampling practically every kimberlite occurrence in southern Africa, had developed a feel for the right terrain for diamonds. Canada, with its vast areas of primitive 'shield' exposed, was undoubtedly prospective, and by the 1960s, as we have seen, a number of glacially derived diamonds had been found and even some kimberlite. Arnold Waters was delighted to be able to sign on his former lieutenant to apply what he had learned in Africa to his home turf.

Some of Tremblay's experience of diamonds, kimberlites and their exploration would stand him in good stead. In Waters's view, sampling stream sediments for indicator minerals should work as well in the wastes of Canada as it had in Africa, so that is what he determined Tremblay would do. Without hundreds of low-cost African samplers to call on, the work would be slower, but access along public roads was good, in southern Canada at least.

Tremblay's teams found indicators in profusion, mostly leading nowhere. Nevertheless, the technique of following trains of kimberlite indicator minerals (KIMs) was fully vindicated when an indicator mineral trail led to a kimberlite pipe in central Ontario. The pipe was barren. First-phase analysis of another pipe, opportunistically discovered, showed it was too low-grade to warrant further exploration. Month after month of routine grid-sampling for KIMs found no El Dorado. The work seemed futile, and in 1965 Tremblay left De Beers. Waters, disappointed, found a replacement and the work continued.

Nothing could have been more daunting than mounting an exploration programme over a country second in size only to Russia – very nearly 10 million square kilometres – most of which is underlain by diamond-prospective craton. Other factors added to the difficulty. Firstly, kimberlite pipes burst through the Earth's carapace with a minimum of disturbance of the ground around them. They are small, mostly not much more than 10 hectares, they come from a depth of well over a hundred kilometres and they come quickly, like a thief in

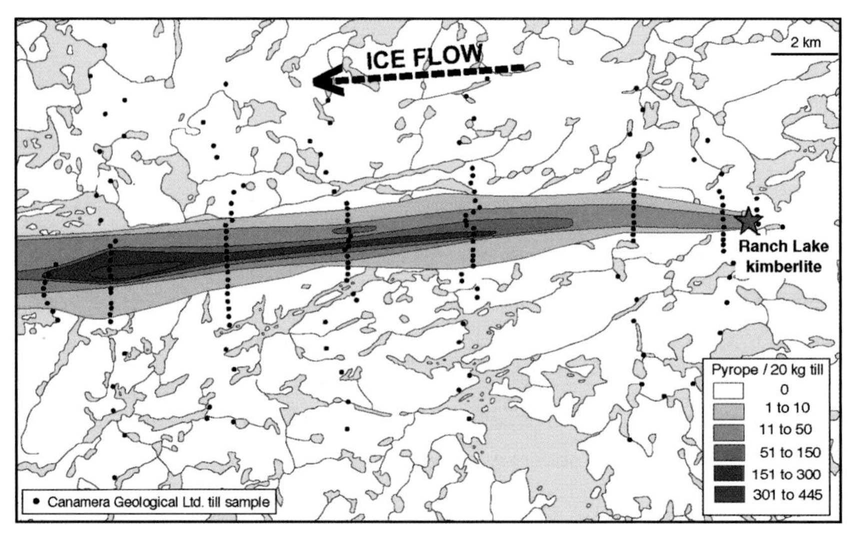

Continental-scale ice sheets can concentrate resistant material like garnets for tens of kilometres, as is evident on this map showing garnet distribution from the Ranch Lake kimberlite near Lac de Gras.[2]

the night. Beyond the perimeter of the pipe there is almost no sign that they're there. To see the evidence involves getting on your hands and knees and looking for mineral grains usually no bigger than a grain of rice. There is practically no other mineral geologists search for that does not have an aureole – a tell-tale envelope of rock – around its deposit to lead you into it, some quite extensive, others more restricted.

Where diamonds were scavenged from their pipes by slow-moving ice sheets, glaciation after glaciation, the search for indicator minerals challenged the exploration teams in a way that nothing in Africa had. Ice sheets spread right over the land, not like streams and rivers, which confine themselves to well-defined valleys with a clear gradient. The ice-borne spread of resistant minerals like diamonds was vast and almost random: the kimberlites had to be there somewhere, but for a long time no clues were given as to the location.

In the early 1960s geophysics was in its infancy, and conducting surveys over large areas using light aircraft much more so. Diamond exploration in Africa had involved covering the ground to collect samples, sometimes by vehicle, mostly on foot. In Canada geologists were looking at millions of square kilometres, quite a lot of it inaccessible for months at a time. But in the end they triumphed, thanks partly to two South Africans. Theirs is a country tailor-made for mineral exploration, and over the decades many South Africans have made it to the top in the international mining and exploration industry.

An airborne geophysical survey being flown in Canada.

We have already met Chris Jennings in the survey of Botswana in the African chapter. To see how he resurfaces in Canada we have to go back to the Kalahari of Botswana, a terrain strikingly different from the Canadian tundra where he would be a participant in the discovery of an exciting new kimberlite field. After a long and useful stint with the Geological Survey of Botswana, Jennings had joined the Canadian company Falconbridge Explorations as chief geologist for its western division and later exploration manager. It was not long before Jennings persuaded his Canadian parent that airborne geophysics in northern, central and southern Botswana would very probably show up kimberlite pipes. The large number of kimberlite pipes revealed by drilling was ample vindication for Jennings, and Falconbridge joined the ranks of diamond explorers.

Falconbridge's parent, Superior Oil, delighted with Jennings's and Falconbridge's success, decided it would join the fray on its own account. It appointed a South African-turned-Australian, Hugo Dummett, a passionate and gifted exploration geologist, to head the venture. Unlike Jennings, who was turned down by De Beers when he applied to join them as a young graduate, Hugo Dummett had worked for them for several years and hated it. Both he and Jennings had reason to want to outflank the diamond giant in their quest for a fertile kimberlite discovery, whether in Botswana or North America.

To observers of the race for the next big diamond mine, it might have seemed that the essentially competitive relationship between Jennings and Dummett must be one of conflict. In fact it was not so: they collaborated, and if there was banter between them it was friendly. Ultimately both answered to Howard Keck, Superior's no-nonsense boss.

To begin with, while Chris Jennings evaluated the Gope cluster of pipes in Botswana, Dummett was Superior's man in North America. Although there were other companies snooping around – and, under cover, De Beers' Arnold E. Waters Jr. – he didn't have to worry about Falconbridge: Chris Jennings had more than enough hay on his fork to keep him busy in Africa. Dummett did an exemplary job of preparation, including reading everything he could find on the US kimberlites, visiting the few outcrops and seeing what diamonds the Smithsonian could show him, and was soon ready to start his search.

But before then, there was the pivotal aspect of sample preparation to attend to. As they collected samples of grit and sand from streams over a wide area where kimberlites were known to occur, for investigation by the leading diamond geochemist, John Gurney in Cape Town, the concentrates of heavy minerals to be sent to Cape Town had to be recovered from the samples, a simple enough process. While the kimberlite indicator minerals – garnets, ilmenites, spinels, perhaps even chrome diopsides – were what they were looking for, and what Gurney would analyse, they made up a very small part, a few per cent at the most, of a sample from the field that weighed a few kilograms. It simply was not practical to air-freight hundreds of kilograms of irrelevant material from Tucson to Cape Town. Dummett needed a basic laboratory where the heavy mineral concentrating could be carried out, and a technician to do it.

Kimberlite indicator minerals (KIMs) and a diamond.

This is where Chuck Fipke enters the story.[3] Fipke had studied at the University of British Columbia with a Superior Oil colleague of Dummett's, the geologist Mike Wolfhard. In 1978 Fipke approached his old classmate with a proposal that his company, C.F. Mineral Research based in Kelowna, British Columbia, undertake sample preparation, for any mineral or metal they liked, on behalf of Superior. Dummett saw Fipke's boundless enthusiasm and knew he would apply his energy to producing the concentrate they needed with all the necessary urgency. Could he collect the samples as well? Dummett asked Fipke. Sure, no problem. Fipke roped in another geological contact, Stew Blusson, to help and, together with some young assistants, the Canadians started sampling Dummett's area of interest in Arkansas.

It was not long before they found a pipe, about a kilometre from the Crater of Diamonds. It was diamond-bearing but very low-grade with stones worth a few dollars a carat. The samplers moved on west, to Colorado and Wyoming. If Fipke, Blusson and the samplers showed grit and determination in their work, confronting rattle-snakes on a regular basis as they walked or crawled through thorny Arkansas undergrowth to reach sampling sites, and irate gun-toting landowners just about as frequently, the geologist Tom McCandless was even more exemplary in his application. He had lost the use of both legs in a motor-bike accident. As far as his fieldwork was concerned, it just slowed him down a bit, having to get around in a wheelchair. But nothing dampened his ardour and he became a valued right-hand man to Dummett.

By now Fipke had been bitten by the diamond bug. So when kimberlite was reported from a cutting through the Rocky Mountains east of Kelowna, he pleaded with Dummett to let him do some prospecting in the area, with Fipke retaining an interest in the venture. Dummett agreed.

The nearest town to where the kimberlite had been exposed was, ironically, named Kimberley, after the South African diamond centre. The highest city in Canada, Kimberley in British Columbia had been a mining town since 1917, though history had thus far not repeated itself and only lead and zinc were mined in the Sullivan Mine outside town. Now the prospectors all hoped that fate would smile on them and would show how apt the name was after all.

All the same, this Kimberley was a far cry from the plains of the Northern Cape. Fipke and Blusson were to find within hours that they had exchanged deadly snakes and furious farmers for precipitous Rocky Mountain valleys and snow-capped peaks, the most hazardous helicopter-flying conditions imaginable. Undeterred, they collected their samples. Soon chrome diopsides were turning up in the laboratory and they were finding the kimberlite pipes from which the minerals were coming – and lots of them – where few people would have dared to fly. They took John Gurney by helicopter to their new kimberlite field.

Gurney had been a Tom Clifford devotee from the beginning and knew that the Rocky Mountains were well off the nearest craton. Not only was he appalled at the perilous terrain they were planning to prospect, but he knew the ground was wrong geologically. He told them so in no uncertain terms. They moved north, deep into the Northwest Territories, to look at a pipe Fipke had heard about from a friend in Vancouver. Although De Beers had walked away from it as being too inaccessible ever to make a mine, Dummett agreed that it should be investigated.

De Beers' Snap Lake Mine (the only underground mine in Canada), northeast of Yellowknife, typifies the countryside in what is aptly called the Barrens.

In the meantime De Beers' Dr Arnold Waters, advised by his one-time North American chief, Mousseau Tremblay, had been searching for the source of the diamonds manifestly brought from the north to where they were found around Lake Michigan and, more recently, in Canada. Shrewd observers noted that the Belgian company exploring for diamonds in the Great Lakes region of the United States had exactly the same directors as De Beers and that the project geologists' accents were more reminiscent of Kimberley than Antwerp or the Congo.

In 1973 Waters had taken a giant leap to the north when he came across a paper published in the *Canadian Journal of Earth Sciences* by two Canadian geologists reporting kimberlite on Somerset Island, well north of Hudson Bay, in fact more than 500 kilometres north of the Arctic Circle. With two generations of olivine, 'chrome-pyrope' and phlogopite mica, the kimberlite described in the article sounded just like what he knew so well from South Africa. Conceivably, this was where the Great Lakes diamonds had started their journey. When the Somerset Island kimberlite tested very low-grade, Waters and his teams moved westward, into and across what are descriptively called the Barrens.

So, coming from opposite parts of southernmost Canada, De Beers and Fipke converged in the Northwest Territories at a locality called Blackwater Lake late in the summer of 1981. Fipke and Blusson found that De Beers must have just left the lakeside prospect when they arrived. They sampled gravel bars lying on bedrock – the perfect place to sample – in creeks around the De Beers camp. Where they found open ground, they staked it.

Soon word came back from McCandless and Dummett. They were as excited as Fipke had ever heard them: the samples sent to Tucson and on to Houston to be analysed were full of the G10 garnets that Gurney had advised were practically a sure-fire indication of diamondiferous kimberlite. Fipke danced a jig. He beefed up the sampling and staking teams. Yet in the autumn of 1981, on the threshold of a major breakthrough, Fipke and his men pulled out. It was no longer safe to work: the days were shortening fast and the temperatures getting dangerously low. At the same time he couldn't wait to get back.

But it had all been too good to be true. As Fipke celebrated the start of 1982 bursting with excitement at what lay ahead and impatient to get started, a bombshell was dropped. Superior Oil announced that it was pulling out of Canada. Although Fipke and Blusson had never been employed by Superior Minerals, their work in the Northwest Territories had been funded by Falconbridge–Superior throughout. It had been Fipke, after all, who had got them into the Northwest Territories in the first place. Recognising this as they wound up their Canadian mineral exploration business, the management of the Falconbridge–Superior Oil joint venture ceded the rights to the two explorers, in fact to a company they had formed called, rather grandly, the Blackwater Group.

Chris Jennings (in red parka) and associate at an exploration project in Northwest Territories, a typical scene.

From 1978, when Dummett started in Arkansas, to the end of 1981, Superior Oil managers and shareholders had seen hope and excitement evaporate, time after time. They had staked Gurney's research, helped Fipke establish his laboratory, and funded an exploration programme in the United States and Canada that thought nothing of hiring expensive helicopters when the need arose. For both Superior Oil and Falconbridge, Canadian diamonds had all but lost their credibility. Dummett, a passionate explorationist, moved on to other mining commodities and other parts of the world. He would be back. Jennings, on the other hand, had diamonds in his blood and had heard enough about developments in the Northwest Territories to know that what was happening there was too good to walk away from. He changed companies but not his dream.

The London-based Selection Trust had been started in 1913 by the American mining engineer Alfred (later Sir Alfred) Chester Beatty. After the First World War, Chester Beatty's entrepreneurial skill saw him grow his creation into a major multinational mining company. By the 1960s it was a force to be reckoned with in the fabulous Northern Rhodesian (now Zambian) Copper Belt and in the exciting young diamond fields of West Africa. Other successful mineral strikes followed, in Australia, the United States and elsewhere. In 1981 the BP group, suddenly mineral-hungry, acquired a controlling interest in the company. In Canada it did not take long for BP–Selection Trust to set their sights on an acquisition of their own: Dr Christopher Jennings for exploration manager. Jennings accepted their offer.

Though Selection Trust (Selco) had more experience in diamonds than any company other than De Beers, it was a multi-commodity company, and Canada was known for base metals first and foremost, not diamonds: copper, lead, zinc, nickel. These were the elephants on the Canadian prairies. The BP–Selco team followed the trail of kimberlitic minerals up ice and eastwards from Blackwater Lake to the edge of a line linking Great Slave and Great Bear lakes. Here they lost the trail.

Equally importantly, Chester Beatty had retired in 1968: the Selection Trust of old, too, was gone, and in its place a mineral exploration company under the watchful eye of an oil-pumping parent. Was history about to repeat itself? Jennings did not have to wait long for the answer. His funding dried up after two years and he moved on. On the face of it his subsequent time with the Canadian junior miner International Corona Resources proved as frustrating as his previous two attempts to find the great Canadian diamond mine. Indirectly, though, his time with Corona was to be the key that would at long last open the way for him into diamond mining.

We have seen the convergence of the ever-determined Fipke and De Beers in the top end of the Northwest Territories. In a move that would surprise many observers and disappoint the in-house proponents of the Canadian search, De Beers was excluded from the next round of prospecting. A strategic rethink in Johannesburg led to an almost total withdrawal of the diamond giant from Canada in late 1981. They had been there for decades without coming close to an important discovery, and this on the back of equally futile and prolonged searches in Brazil and Australia. On the upside they had two major mines in Botswana at an early stage of production, so their dominance of the world diamond market was assured for years to come.

In the meantime Fipke had progressed from the Blackwater Group, which had inherited Superior's claims, to list Dia Met Minerals on the Vancouver Stock Exchange, which would be the vehicle for his endeavours well into the future. Fipke would spend seven years following indicator trails and raising the necessary funding for high-cost exploration in one of the least accessible and most hostile environments imaginable. It is a tribute to him that through those long hard years he never lost his focus or the belief in his kimberlite pipe shedding diamonds across the Barrens.

After many false leads, his search had taken him to Lac de Gras, 300 kilometres northeast of the Northwest Territories capital, Yellowknife. The French for 'fat lake', the name comes from Ekati, given it by the local Dene Indians, and is thought to derive either from the quartz veins running through the granite outcrops on the lake shore, which reminded them of the marbling of fat in caribou meat, or because they associated the place with the smell

A Canadian diamond exploration camp in the winter.

of burning caribou fat from the days of camping at the lake after hunting. Today few people think of much other than diamonds when they hear the names Lac de Gras or Ekati.

Fipke followed a steadily gathering trail of indicator minerals till he was at Pointe Lake, just north of Lac de Gras. He knew he was close. It was time to protect the area. Wary of setting off a rush, he filed his claim block under a nominee identity. His stratagem failed, though, and his cover was soon blown.

Chris Jennings, still with Corona, realised that a huge post-glacial lake had connected the Great Bear and Great Slave lakes. The muds deposited on the floor of this lake had buried the kimberlitic minerals, and the probable source of these minerals would lie up-ice, further to the east. While he was aware of Fipke's eastward thrust for the source of the Blackwater minerals, he was fairly sure that Fipke was not aware of Corona trying to leap-frog him to the source. Jennings closed in from the east, using the field geologist Leni Keogh. With company funding, 'borrowed' unauthorised from another project, behind them, their pace was faster than Fipke's and they were at Lac de Gras quite soon. Keogh found signs of samples having been taken. Her sampling pace became more urgent.

When Jennings joined her, they found microdiamonds in a sample, apart from dozens of garnets and chrome diopsides. The mineral chemistry of the garnets screamed a good diamond source. When Jennings requested $100 000 from the Corona board to stake

CHRIS JENNINGS

Read any book about Canadian diamond history and you will see reference, commonly frequent, to Chris Jennings. Generally they will recall Chris's beginnings in Botswana, where in the 1970s he found the first kimberlite pipes after the De Beers' discoveries. They may not cover the discovery by the company he formed, SouthernEra, of the fabulously high-grade Marsfontein pipe and Klipspringer kimberlite dyke in South Africa: these came after Canada. Chris is humble enough to stress that he could not have achieved what he did without assistance from other geologists and, particularly in the early stages, the pivotal involvement of John Gurney with his then new techniques in mineral geochemistry. But Chris, backed by wife Jeanne, was the driving force. Mostly the companies whose exploration he headed did not have the staying power and in the end he had to start his own; but he never gave up. Ultimately he was duly and hugely rewarded, with the Diavik Mine in Canada and the South African mines closely linked with his name.

Geologist Eira Thomas holds a length of drill core with a diamond at the end, from a hole she sited on an island in Lac de Gras to test a hunch that there was a kimberlite pipe under the lake. It happened in April 1994 when Eira was the 25-year-old leader of the field team in the area of Aber Resources, the company her father Gren had founded. The discovery would lead to Aber bringing Rio Tinto in to build the Diavik mine and would make Eira a legend in the global diamond business.

claims over the area, he was turned down despite a letter written in July 1991 predicting the discovery of economic pipes and a huge staking rush. Frustrated, Jennings once again moved on. He would go it alone.

Fipke, meanwhile, had enlisted the help of his old friend and ally, Hugo Dummett, now with BHP and a generous budget. Fipke suspected that the perfectly round lakes, roughly the size of many of the kimberlite pipes he had read about, were reflecting exactly that. He argued, correctly, that the kimberlite, being much more susceptible to weathering and glacial erosion than the surrounding granite, would have formed depressions, then lakes. Having seen the chemistry, Dummett fought long and hard with BHP management before they agreed on a joint venture with Dia Met. Without the persuasive powers and determination of Dummett, this would not have happened. They eventually backed his hunch and decided to drill Pointe Lake. As drilling got under way, airborne geophysics confirmed that practically all the small lakes were very probably over kimberlite pipes.

On 9 September 1991, the drill, boring an angled hole from near the edge of the lake into the rock under it, hit kimberlite. John Gurney, fortuitously in Canada to attend a diamond symposium, was called to see it: he couldn't have liked it more. Laboratory tests would soon confirm his optimism. Dia Met had found Canada's first diamondiferous kimberlite pipe. Fipke's remarkable resolve, grit and single-mindedness had been rewarded.

Try as they might, BHP could not contain the news of the discovery nor did they have the freedom to claim much more ground than the vast tracts they already controlled. It was not long before Chris Jennings was back in Yellowknife, this time at the behest of two Toronto-based exploration entrepreneurs, Gren Thomas and Bob Gannicott. Jennings knew exactly where Leni Keogh's mineralised samples had come from, just north of BHP's claim block. Jennings would claim it for the Toronto principals. When they could send a helicopter into the area in the spring of 1992 they found they were too late: De Beers had already claimed. The South African giant was back. Undeterred, Jennings decided he would claim 'up-ice'. He knew from his days in Botswana that a kimberlite field may extend over tens of thousands of square kilometres, at over 240 acres (100 hectares) per square kilometre.

The Dia Met–BHP joint venture had staked 850 000 acres, De Beers over a million, and the new Toronto company, Aber Resources, had well over half a million. In all, about 100 square kilometres had been claimed by the early summer. It was not over, though.

Until then Jennings had worked on behalf of the Aber principals, with a royalty on future production written into his contract. At this stage he decided the time had come to tie up some ground for himself. He bought a shell company on the Toronto Stock Exchange for next to nothing, and soon Southern Era's name showed up on the claims maps for the first time, with a holding of 700 000 acres. Before this, so as not to be left out of the greatest staking rush in history, Rio Tinto had teamed up with Aber.

Diavik in the summer ...

Chris Jennings recalls those days: 'Here again, we have another unsung hero, a bit like Hugo, who fought so hard to have BHP partner Dia Met at Ekati. In this case it was John Collier, Rio Tinto's head of exploration … Like Hugo, John managed to persuade Rio Tinto to accept his recommendation to joint venture with Aber ...

'After helping Gren [Thomas] and his team to stake their initial claims in NWT in November and December 1991, I decided to take a break in South Africa and Jeanne [Chris's wife] and I booked to fly via New York. At about this stage, Aber was running out of money to stake new ground and they targeted Rio as a potential partner. Bob Gannicott had just returned from seeing John Collier in London about some non-diamond venture in Greenland. During their meeting, Bob was asked if he knew anything about the diamond staking rush in Canada's NWT. Bob's reply was: "Speak to Jennings." As a result and after much hesitation on my part, Rio Tinto persuaded us to fly via London. We arrived on a Saturday morning and were promptly driven to John's house in a London suburb. After explaining my involvement in the predictive power of mineral chemistry, I stated that the Lac de Gras chemistry was better than many economic pipes which I had in my database.

'I also told him that I felt that the Dia Met–BHP (now BHPB) [joint venture] probably did not contain all the prospective ground. John was very excited at this news and, soon after, Rio Tinto signed a deal with Aber.' The association between Aber and Rio Tinto would lead to the opening of the Diavik mine.

... and in the winter.

By the summer of 1993 the first phase of the rush was over. BHP had scores of geophysical targets which they were testing, many of them proving to be due to kimberlite pipes. The other explorers, too, were consolidating. That the first-comers were busy entrenching themselves did not by any means stop wave after wave of new arrivals from seeing where they could find an untenanted corner, or ground on the edge of the first blocks, or even an existing claimant who was looking for investment. The stock exchanges in Vancouver and Toronto buzzed as never before; so too the telegraph lines from Yellowknife. Hotels and motels in the capital were bursting at the seams, aircraft charter companies could not find enough craft, and hardware stores failed to keep up with the demand for staking and camping material. It was chaos, activity and hype pitched at unprecedented levels. Flaws were found in hastily drawn-up partnership agreements: old friends sued each other.

For some the publicity was followed by disappointment. In a number of prospects diamond grades were not nearly so good as anticipated: certain mines were first less probable and then written off. There was some good wheat, but plenty of chaff. All three main front-runners – Dia Met–BHP, De Beers and Aber–Rio Tinto – wound up with several good mines each in their claim blocks around Lac de Gras. They continued with airborne geophysical surveys knowing, for the first time, that the Canadian shield was as prospective as any. As they had hoped, there were other diamond fields to be discovered, not just Lac de Gras. The surveys turned up kimberlite pipes that are big enough and rich enough to be turned to account in more hospitable parts of the country, such as Ontario and Saskatchewan. De Beers is there as a major force, but so for the first time are other top-ranked mining companies with world-class mines, BHP Billiton and Rio Tinto. In a world that is increasingly prosperous and hungry for diamonds, the other mining giants are selling their wares outside the De Beers marketing agency, the Diamond Trading Company.

Canada is the world's sixth most important producer in terms of carats produced, though it jumps to third place in value of production, just behind Russia and a little ahead of South Africa. The disparity between size and value of production can be attributed to the value of Canadian kimberlite: more than half of the world's mines in the grouping with rock value between \$100 and \$1000 per tonne (the top category) are Canadian. The grades (in carats per tonne) of Canadian mines are high, as is the dollar-per-carat value of the stones. Unlike many mines in southern Africa, for example, the stones are consistently gem-quality, and so the mines in Canada do not have to depend on occasional very big, exceptionally good-quality diamonds to be payable. By the same token, diamonds considered big by South African standards are unknown. In July 2008 a 25-carat stone found at the Gahcho Kue prospect in the Northwest Territories was billed on the Internet as the largest ever found in Canada. What makes the stone even more remarkable than its exceptional size is that it was recovered from drill-core, which, after all, represents a tiny fraction of the body being drilled. To put the size of Canadian diamonds in a global perspective, it would probably not be an exaggeration to say that 25-carat stones are found daily in different parts of Africa. Having said that, the deposits vary as kimberlite pipes will vary wherever they are found: in size, grade, value of stones and age. They have provided a wonderfully fertile field for geologists and geochemists and will continue to do so.

A rough Canadian diamond on kimberlite chips.

What of the geologists who played such central roles in the discoveries that put Canada on the diamond map? The most direct participants, Chuck Fipke, Stew Blusson, Hugo Dummett and Chris Jennings, all did well financially out of their remarkable resolve and unwavering tenacity even if it did not bring them all personal peace in the end. Dummett, who received a number of awards for his contribution to the Canadian diamond discoveries, died tragically in a motor-car accident in 2002. The others, too, were recognised with a variety of awards, Jennings scooping the 2007 Lifetime Achievement Award, presented annually in London at the *Mining Journal*-hosted Mines and Money meeting. He was also honoured with an award by the Prospectors and Developers Association of Canada (PDAC), while Fipke received the highest award from the PDAC, and Blusson was similarly honoured by the Geological Survey of Canada.

IN CANADA a handful of dedicated geologists had occupied centre stage in one of the great exploration dramas of the twentieth century. Their decades of courage and unshakable conviction were generously rewarded and the country's diamond representatives will continue to attend international conferences with heads held high. In the United States, the Crater of Diamonds in Arkansas still welcomes hopeful prospectors and their families: a lucky few go home with a stone that they will treasure but probably never have cut, most with hopes dashed. Nature is fickle in her favours.

An aerial view of the Argyle Mine in the Kimberley, Western Australia.

CHAPTER TEN

AUSTRALIA

For well over a hundred years diamonds were known in Australia, like Russia, before the discovery of the huge deposit that would put the country on the world map as an important producer. In all of the biggest producing regions of the world after Brazil – South Africa, Russia, North America and Australia – the first stones were found in the middle part of the nineteenth century. While in South Africa workable concentrations of large gems were turned up soon after the first diamond had been identified and, two years later, the mother lode, in New South Wales, by contrast, where a diamond was reported as early as 1851, the mother lode has never been located, nor were the alluvial deposits there of any particular importance in the global scheme of things. All the same, they were enormously significant to the finders and curious geologically, so we will give them their due.

Unlike the discovery in Western Australia over a hundred years later, the New South Wales diamonds were found by accident. In that respect they were like the first diamonds discovered in most parts of the world in those days. Mostly they were found in gold pans or sluices: in New South Wales tin was being recovered from so-called deep leads – actually palaeo-river gravels – on quite a large scale, and diamonds rolling off the end of the sluices used for concentrating the tin were picked up by children playing while their fathers worked. They were shiny and made pretty playthings.

For the old-timers diamonds were a speculative prize to be looking for. Not only were they few and far between, but the stones were small. Unlike tin or gold, whose price was fixed and for which a prospector had a good idea of what he would earn, he had little inkling of what the merchant would pay for his diamonds. So it was only thanks to an unusually observant prospector that the first diamond was noticed among the camp children's baubles. Though history doesn't relate when diamonds became the primary objective of prospecting, and not just a byproduct of tin or gold, suffice it to say that by the 1870s, when discoveries in the Cape Colony began to fire imaginations around the world, diamonds were being mined at a number of sites in New England, in

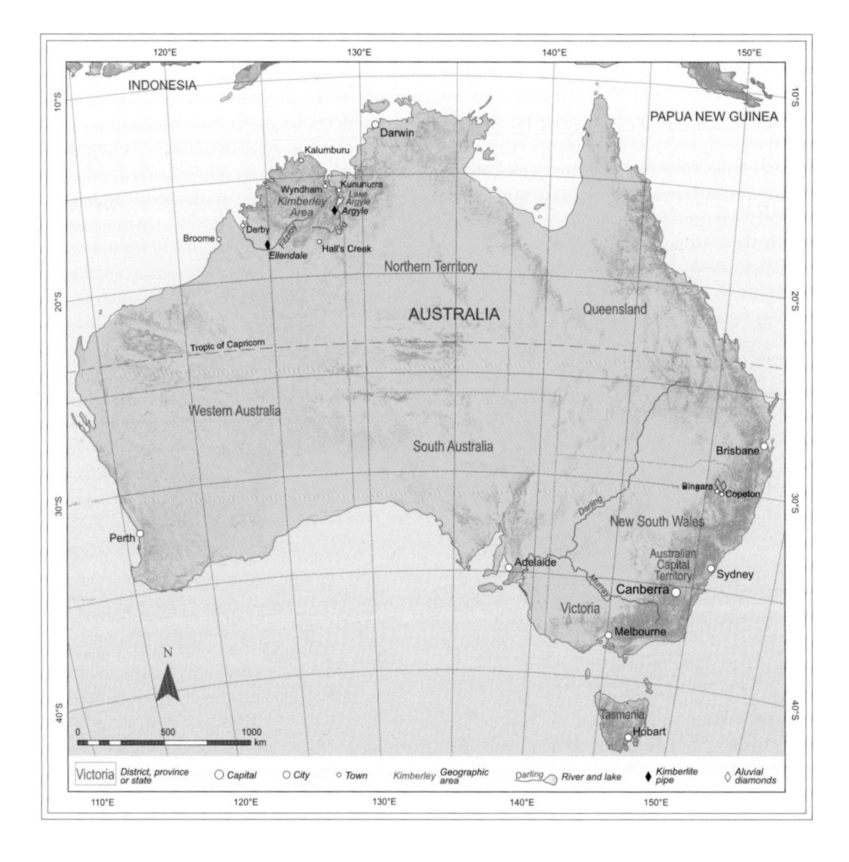

northeastern New South Wales. The most important fields were at Copeton and Bingara, a little over 200 kilometres from the coast and about 400 kilometres north of Sydney.

The larger of the two New England fields was Copeton, where from the 1870s to 1915 about 300 000 carats are thought to have been produced from some 25-odd workings. Of these the Star of the South was probably the biggest, and the field produced in its best year (1899) 25 477 carats. Considering the gravel was reached by a 27-metre-deep shaft, that the stones averaged 'between 3 and 4 per carat', with the biggest stone 6.25 carats, and that difficulty was experienced in finding a market for the Copeton diamonds, this is a remarkable accomplishment, even if it pales to nothing by comparison with the 34.4 million carats produced from Western Australia's AK1 pipe at the Argyle mine in the year 1989.

On the plus side the grade of the 'wash' was, compared with any other gravel, extraordinarily high, at 3 to 4 carats per tonne. Furthermore, there was enough tin to play a large compensatory role for unglamorous or difficult stones, roughly 4 kg of cassiterite (SnO_2 or tin oxide) per tonne of wash.

Bingara, mined from the mid-1870s to about 1902, was a substantially less productive field, with even smaller diamonds (from 5 per carat down to 20 per carat) than Copeton, though apparently of a better quality. Like the Copeton diamonds, the Bingara stones were notoriously hard, making them difficult to cut. Tin was a rare byproduct, though gold was an occasional sweetener, more so than in the Copeton workings.

Other deposits had been, and continued to be, found further south and closer to Sydney but, together with Bingara, they didn't add up to what was recovered from Copeton. It is not difficult to understand why few people thought of Australia as a diamond-producing country until 30 years ago.

Source of the New England diamonds

Copeton diamonds: not unusual to look at, except to experts and (especially) manufacturers.

Where, everyone asked, are the kimberlite pipes in New South Wales? Some, like the De Beers subsidiary Stockdale Prospecting, went beyond idly wondering and prospected long and hard for them, without success. When diamonds were found in Western Australia in the 1970s in lamproite, a rock really quite different from kimberlite, the hopes of finding a 'para-kimberlitic' mother lode in the east were revived. By now, though, Clifford's rule about diamonds occurring only in kimberlites and allied rocks piercing cratonic crust had become gospel, and New England's 'off-craton' position was problematic in this model. In any event Stockdale – and others – found no kimberlite. In a quaintly curious turn of phrase, the New South Wales deep leads, and others like them in Southeast Asia, have become known as 'headless placer deposits', and their primary source remains mysterious and resolutely elusive.

Might they have come from a rock unrelated to kimberlite? people asked. Geologists set about answering the question by studying the crystal form and internal structure of the diamonds as well as the chemistry of their inclusions. An accepted method of learning about the environment of the diamonds' formation, this is usually a fertile field of research. The work led to some answers, and more questions. The diamonds were found to fall into two broad categories, called by the researchers Group A and Group B. Group A were ancient, deep-seated mantle diamonds: were they formed under Antarctica before Gondwana broke up and river-borne, or even glacier-borne, to their present position by Gondwana rivers? It is a popular theory. Group B diamonds, unlike any others known, were probably formed much later, as Australia drifted eastwards over subducting Pacific plates and then brought to the surface by some magma other than kimberlite. It is a mechanism so unusual, so untested, that it must stand as a question until more work is done and more clues garnered.

If kimberlites remained elusive in places with evocative names like New England, far to the west – in a very different part of the country – pioneer geologists wondered if they were getting warm. In 1940 two of the great names of Australian geology, Arthur Wade and Rex Prider, published a paper on the findings of their fieldwork in the West Kimberley area. Wade had spent two years mapping the oil-prospective Canning Basin in coastal northwestern Australia during the mid-1930s. While in the field he sampled curious-looking intrusive plugs which made conspicuous prominences in the mostly flat, arid landscape. He took his samples to the young petrologist Rex Prider at the University of Western Australia in Perth, who recalled: 'what an unimpressive lot they were – mostly weathered, difficult to section … and study. However, I took this collection to Britain in August 1936 … where they constituted part of my PhD research. It was during this study at Cambridge that the lamproitic nature of the West Kimberley rocks was recognised and a mantle origin and possible relation to kimberlite were postulated. The results of this work were published in the *Quarterly Journal of the Geological Society of London* in 1940. There were many, I believe, who thought my ideas on the origin of these rocks were crazy, but I always said: "If they find a diamond in these rocks it will prove I was right, and if they don't find a diamond in them it won't prove I was wrong."'[1]

At the time, there was a war on, and gems and jewellery were the last thing on anyone's mind. A hypothesis from a petroleum geologist and a university petrologist about rocks in a remote part of Australia was not accorded much gravitas by the diamond industry. Wade went from strength to strength in the oil business, Prider returned to his leucite lamproites and the Kimberley region slipped back into dusty obscurity where it stayed for nearly three

The Kimberley outback.

decades – until, in a very different time, someone remembered a geological paper about Western Australia that had mentioned the word kimberlite.

◆◆◆◆◆

THE MIDDLE DECADES of the twentieth century saw a drastic shrinking of the globe and multinational mining companies came into their own: RTZ, Newmont Mining, Falconbridge, International Nickel, Phelps Dodge and others. The world was their mining camp. Nowhere was too remote for them, and new search technology and geological understanding were added to inventories daily, or so it sometimes seemed. Geologists were eager to be in the field: reputations, even good money, were there for the making.

As previously parochial mining and exploration companies went multinational in these decades, a parallel development was the notion of mining company joint ventures. In South Africa the great success of the newly opened Palabora copper deposit, attributable to powerful synergies between London-based RTZ and America's Newmont Mining, was just one example. Not only was there the obvious advantage of risk-spreading; it was commonly found that different companies contributed different skills and approaches that, though diametrically opposed in some cases, were not necessarily incompatible.

In the 1970s Australia became a globally important player in the mining business. Broken Hill Proprietary (BHP) and the Rio Tinto daughter company CRA (Conzinc Rio Tinto Australia) were world-class mining companies: no geologically sound concept was out of bounds. There were rumours of diamond activity in northern Canada, where De Beers had apparently not won the race for the first mine. So the monolithic diamond giant was fallible after all. Had they erred in Western Australia? some wondered.

Ewen Tyler can look back on a lifetime of extraordinary achievement, and satisfaction.

Even before the explosion at mid-century, there was a stirring that foreshadowed the big bang. A family that exemplifies the transition from the embryonic stage of the multinational mining fraternity to its vigorous youth is the Tylers. 'Watty' Tyler graduated from Sheffield University with a master's degree in geology in 1924. After a decade in Burma and various Central African countries, he was sent by RTZ to Australia. From there he went to Sarawak, on the island of Borneo, to prospect for mercury, which was needed for military detonators. His sons, one of whom was Ewen, and their mother joined him there in August 1940. It was here, at the age of 12, that Ewen found his first diamond.

After the war Ewen enrolled at the University of Western Australia, intending to major in chemistry but later switching to geology. It is testimony to his strong entrepreneurial streak that he had not even graduated before he was branching out on his own to create his own exploration company. Even if he didn't find a mine, Ewen got his first taste of the thrill of being where no one had ventured before, looking for his El Dorado.

In 1950, recently graduated, he decided he would visit his father's old hunting grounds in Africa. Watty had told romantic stories of gold mining at Geita in the far north of Tanganyika. Sixty years on, Geita is Tanzania's largest producer of gold and the gold giant AngloGold Ashanti operates a huge open-pit mine there. When Ewen arrived, the mine was a small underground operation, in a remote corner of Africa with practically no infrastructure. But the eight years Ewen spent there gave him a breadth and depth of experience in mining and exploration – not to mention managing hugely challenging logistics – that he had never dreamt of.

In the late 1950s Ewen's contract in Africa ended and Geita's parent, Tanganyika Holdings, took him on in London as a technical consultant. For ten years he monitored its operations in Africa, mainly the Katanga copper mines. After the Congo achieved independence, the mines were nationalised and Union Minière and Tanganyika Concessions found themselves stripped of their assets. They disposed of their stockpile of copper held outside the country and prepared to ply their trade elsewhere, preferably outside Africa. Ewen, now a director, was allowed to set up Tanganyika Concessions in Australia to look for commodities like platinum and diamonds and relocated there in 1969. It was thanks to Ewen Tyler that Tanganyika Concessions, otherwise known as Tanks, became the founder of the joint venture that launched the diamond search in the Kimberley region.

We have seen how Tanks got to Australia. Let us look briefly at how a London-based company called Tanganyika Concessions became a force to be reckoned with in the Belgian Congo. Just over a hundred years ago, rich deposits of copper and cobalt had been found in an area covering thousands of square kilometres in the so-called Congo Pedicle, a limb of that country's Katanga province jutting into Zambia. Recognising that he had no pool of

Not diamonds in Australia but copper in the Congo, where Tanganyika Concessions acquired a taste for mining on the biggest scale.

mining expertise to match that of the British, the ever-ambitious King Leopold II invited Tanganyika Concessions to join the Belgians in the company formed to mine the copper, Union Minière du Haut Katanga. Within a short space of time Tanks emerged as a powerful force in the Congo and beyond.

IN 1969 EWEN TYLER made the first moves to take Tanks into what became the five-party Kalumburu joint venture, which he managed. His objective was very simple: to find a diamond deposit in Western Australia. Tyler thought that the elephant country would be in the north, in the Kimberley, where the geology was most similar to the central part of South Africa where diamondiferous kimberlites had been found. More than that, he knew that this was where his old professor, Prider, had found intrusive rocks which he believed were related to kimberlite.

That he could make a case for what might to some have seemed an outlandishly speculative venture was characteristic of the man. From as early as he could remember his father had imbued him with his passion for exploring, in every sense of the word: being in new places, where no one had been before, testing new ideas and new techniques, finding mineral deposits that had never been dreamt of, testing his own limits of endurance. For Watty and for Ewen, nothing on earth could be more exciting, more rewarding than mineral exploration.

Ewen hadn't been exploring in Western Australia for long before he found out that an Australian junior company, Arcadia Minerals, held licences over the bodies his ex-professor had sampled years before. His own knowledge of kimberlite and diamond geology was minimal: he would need a consultant to accompany his Perth geologist, Mick Paltridge (also not a diamond specialist), into the field to see the properties. Not only was Chris Smith in the right place at the right time, he was very much the right man, as time would show. Having worked for eight years for De Beers in Africa, he was well acquainted with diamond exploration in tropical terrain, and he'd been in Western Australia for a few years, so this was now his home turf. He liked what he saw in the field – bodies of leucite lamproite – but he knew they would not themselves carry diamonds. Though he had hoped to see olivine in the rocks, there was none. Nevertheless, the lamproites had come from the right part of the mantle, and perhaps the olivine-bearing rocks were present but just not outcropping. So he advised against going in with Arcadia Minerals, but suggested instead that Tanks should consider the Kimberley Cratonic Block as a regional target and explore for kimberlites across it.

They knew the logistics were going to be tough. The Kimberley is some 600 kilometres across, and access to it was then to be gained by one dirt track through the centre from south to north, and a rough track from Gibb River in the centre to the port of Wyndham in the east. Other than this there was practically nothing, in country full of rocky gorges and cliffs: the entire venture would have to depend on support from the air.

Early Western Australian diamonds

A story from the Kimberley that had its origins in the early days of exploration but only later came to light reaches us from a Sydney jeweller. Jules Joris, scion of an old Antwerp family of cutters and polishers and in his time probably the most respected figure in the Australian diamond manufacturing industry, had been called on by a man named Watson, who described a locality in the northern corner of Western Australia where he claimed to have found 'plenty of diamonds'. It would seem that Watson gave Joris one of the stones he had found, a yellow diamond from the Drysdale region, south of the Kalumburu mission station. Joris wrote to one of the monks describing the place as recounted to him by Watson. The monks, who have Joris's letter in their records, did their best to locate Watson's source but with no success.

Ewen Tyler called the joint venture that he started to assemble in 1969 the Kalumburu Joint Venture, or KJV. Did the name owe its origin to the Watson–Joris story? I asked Ewen. Emphatically not, he said, elaborating: 'A precursor to diamond search in the Kimberleys was a copper search, mapping a copper-bearing shale which we were able to trace round much of the Kimberley Block. In this programme we had contact with Father Seraphim Sans at the Benedictine Mission at Kalumburu, so when we were looking for an obscure name, Kalumburu was an obvious choice for what was a clandestine activity for more than six years.

'One of the parties to the KJV obtained the Joris letter: it referred to the Sir Frederick Hills, Mt Hann and the "Vulcan". The Sir Frederick Hills are close to the mission, but Mt Hann would be 200 kilometres to the SSW. We have all visited Mt Hann (Chris Smith many times). While we now know there is a spread of diamonds in the North Kimberley, Mt Hann seemed very deficient! The letter came into my hands long after the naming of KJV, but it attracted our attention. When Father Sans was apprised of the contents of the letter, addressed to a much earlier mission superior (circa 1973), he said, "Oh!, we know all about the diamonds, but we buried them under the concrete floor of the Mission!" If I remember rightly, the letter was to Father Thomas [Gill], who was at the mission from 1926 until he was killed in 1943 in a Japanese air-raid. Father Sans first arrived in 1939 as a young monk. He turned 90 this year [2009]!'

By 1972 fieldwork could start in earnest. In an uncharted and inaccessible region the exploration was going to need to be helicopter-supported, and a laboratory would have to be set up in Perth. Tyler must have congratulated himself on his decision to take the ever-positive Chris Smith onto the books as full-time chief geologist: he was able to give him the moral support he needed, aside from making crucial geological input.

According to Chris Smith, 'We negotiated with a Japanese pearling station at Kuri Bay to service the northwest (there was no road or airstrip; it was supplied from the sea). We would put out fuel dumps along the tracks to extend helicopter range. But there were large 200-kilometre sections without any access, especially in the west, that were going to be a problem and would require fuel drums to be airlifted in by helicopter.

'The helicopter gave you an ideal observation platform to pick potentially good trap sites from the air. However, the rugged terrain meant that you could not always land next to them. Lots of walking and even cliff climbing were still required to get to them and carry heavy samples back out to the chopper.

'We could put our field team of four (Maureen, myself and the two fieldies) together with the pilot into the helicopter. One geo and one fieldie would be put down at the first trap site; the other two would then go on to be put down at the second site. The chopper would return to the first party, who would use a signalling mirror to ensure the pilot saw them, and they would be moved on to a third site when the chopper would return to the second pair and the leapfrogging system would continue.'

With infrastructure all but nonexistent forty years ago, the diamond project had to be helicopter-supported for most of its work.

It took a special kind of person to choose to work in this environment. Maureen Muggeridge, whom Chris refers to, was one. The daughter of the journalist Malcolm Muggeridge, she had just recently graduated from St Andrews University in Scotland and came out to Australia looking for work that was to her liking. Sighting an advert for a geologist in the West Australian newspaper, she came to the KJV for an interview. She had a taste for adventure and insisted that she would not be daunted by the rugged and isolated nature of the Kimberley. Once she was employed, one of her most important qualities would prove to be the painstaking thoroughness she had inherited from her German mother. Her employers were confident that, regardless of terrain, the best sample was always collected. Explorers know that this sort of attention to detail can mean the difference between finding a deposit and missing it.

The team would 'fly-camp' at a logistically favourable site and radiate from there to collect their samples. Chris Smith describes some of them: 'One of the most interesting campsites was at the Japanese-run Kuri Bay pearling station. This was an isolated coastal settlement managed by Japanese and supported entirely from the sea. There was no airstrip. The management and divers came straight out from Japan and spoke little English; the labouring workforce was Torres Strait islanders. There was one Australian only, the wireless operator. They were all very hospitable, showed us how they dived for oysters and then seeded them with grit around which a pearl would form after the oyster was suspended in the sea beneath a raft. Very kindly, they allowed our helicopter fuel and camping gear to be brought in on their supply ship from Broome.

Geologists, and their wives, would have been startled to come round a corner and meet a goanna.

'The Durack River campsite, near where the Gibb River-to-Wyndham track crossed the river, was beside a large, beautiful river pool in which we would gratefully swim at the end of each hot day. Then we noticed it was full of crocodiles, but they didn't seem to mind us and nor did we them. We received a camp cook, a young Welshman who had just arrived in Australia. In the morning we all flew off in the chopper, leaving the cook to settle in. He was horrified to see an enormous reptile several feet long coming into his kitchen tent; in terror he fled to his own tent and hid himself inside, where we found him cowering on our return later in the day. From his description we realised it was one of the enormous goannas (monitor lizards) that lived there, which we would occasionally feed with some food scraps; they are quite harmless. But there was no meal ready for us and the cook took some convincing about his safety before he nervously emerged, still half believing that it was a crocodile that had come for him.'

But this was no camping holiday or adventure tourism experience: it was work, and day after day the teams would be dropped off in inaccessible spots. They would have to walk, climb, dig and then hoist heavy samples into the helicopter. With time of the essence as the helicopter hours clocked up, the sampling teams decided not to process the samples in the field that would contain the indicator minerals they were looking for but send them to Perth as they were. In Perth the concentrating of the heavy minerals would be carried out by a lab belonging to one of the partners and specifically designed for another heavy mineral project. Both Ewen and Chris knew that at least as important as correct collection of the sample in the field was the investigation of the heavy mineral concentrate. With his extensive African diamond experience, Chris would look after this, at least to start with.

He set about training Maureen and another geologist, Robert Mosig, seconded to the project as a matter of urgency. There were a lot of samples to look through, each to be scrutinised with unfailing concentration. They knew that De Beers had covered the ground years before and had left, unrewarded: perhaps they had overlooked vital clues. They could not run this risk.

Using a few microscopes set up in the kitchen of the Tanks house in Perth, they set about their task. Over the years Chris had assembled a comprehensive suite of indicator minerals from Malawi, Zambia and South Africa. He made sure the others familiarised themselves with what the little grains looked like until they could see them with their eyes closed. Bells must ring immediately if anything reminiscent showed up in the Kimberley samples. They should not necessarily expect to find the garnets or ilmenites that are such a sure-fire indicator in South Africa: if they were there, De Beers would have found them.

Chris had noticed a common indicator in South African kimberlites, particularly some in the Orange Free State, that De Beers paid little heed to: chromite, sometimes known by its family name, spinel. His insistence that they should keep an eye open for chromite was ultimately to prove crucial in finding Argyle.

THE TEAM HAD their first sniff of encouragement when, on the eve of Ewen's 45th birthday, 23 August 1973, Chris reported the discovery of the first tiny diamond. Hear it in his words: 'In August 1973, Robert Mosig was combing his way through a concentrate sample from Pteropus Creek (we named the creek informally after the genus of large-sized fruit bat or "flying fox" that was abundant in the trees along the stream bank there). Robert drew my attention to some strange-looking "garnets" in the sample. I looked: they weren't garnets at all; they were large white to pale brown to pink-coloured megacrystal zircons which crystallise in the low velocity zone at the base of the lithospheric mantle. Then my heart leaped as I noticed a shiny diamond in the sample. Also I could see some grains of pyrope, picroilmenite and chromite!

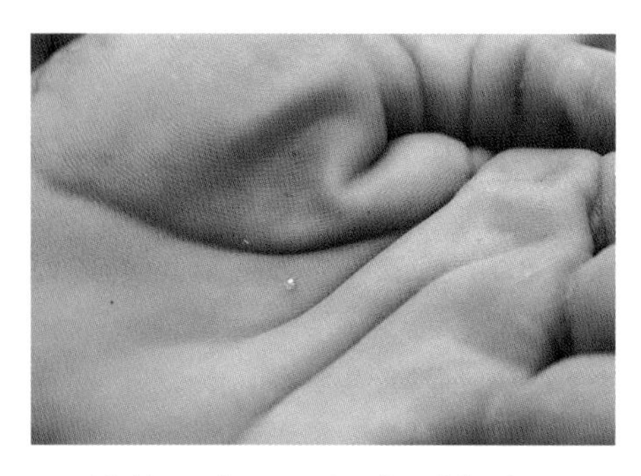

No bigger than a grain of sand, but how important that first diamond.

'I telephoned Ewen Tyler and said, "Ewen, we've found the lot, pyrope, picroilmenite, chromite, kimberlitic zircon and … a diamond!" There was a long pause. Then he replied, "Are you sure?" He then asked me to find a way of testing and confirming the diamond. I said I knew it was definitely a diamond, but he insisted the joint venturers would require separate proof.' In the end Chris convinced them – and check samples from Pteropus Creek contained yet more diamonds. That day, 23 August 1973, is a red-letter day in the history of Australian diamonds.

Meanwhile, the processing of other reconnaissance-phase samples proceeded apace. A sample from a locality called Big Spring, in the South Kimberley, yielded chromite, chrome diopside and diamond: its follow-up would lead to the discovery of Australia's first diamond-bearing olivine lamproite. Ilmenites in other samples confirmed kimberlitic sources in the East Kimberley, too.

A view of the Cambridge Gulf where the Robert Williams *started her maiden voyage for the joint venture.*

CHRIS SMITH

Few companies owe as much to one man as Rio Tinto to their diamond and kimberlite guru, Chris Smith. Arriving in Australia with a useful background in diamonds thanks to De Beers, Chris joined the Kalumburu Joint Venture before Conzinc Rio Tinto Australia had farmed in to the project and long before the first diamonds were discovered. He master-minded critical aspects of the joint venture, both in the laboratory and in the field, and when Argyle had been discovered and the mine established, he moved on. He proceeded to apply his expertise to Rio Tinto's diamond exploration in Zimbabwe and then in India. In the former the company has become the pre-eminent diamond miner, at Murowa, and in India the Bunder Mine will be the country's main producer when it is up and running. He has published dozens of papers, has been honored with the prestigious Gibb-Maitland Award by the Geological Society of Australia, and is counted among the most important contributors to the advancement of diamond geology.

At this stage the decision was taken to build a full-scale laboratory for the venture. Wilf Jones had joined the joint venture from Selection Trust. Using the experience he had gained in their London laboratory, Wilf went on to run a state-of-the-art diamond laboratory in Perth, the brainchild of Chris Smith and others, and built under the watchful eye of the ever-practical Mick Paltridge. It would give the KJV and its successors a significant edge over any competitors in their future exploration.

The remoteness of the places where first-phase sampling pointed to follow-up continued to pose problems. Pteropus Creek, clearly a priority target, was as remote as any but offered the putative advantage of being near the coast. It could be accessed from the sea, particularly as the Berkeley River, of which it was a tributary, appeared navigable for 25 kilometres from the coast. A second-hand harbour boat was bought and transported to the port of Wyndham, at the head of the 80-kilometre-long Cambridge Gulf in the northeast corner of the Kimberley Block. Chris tells us that the boat was renamed the *Robert Williams* after the founder of Tanganyika Concessions, who had mounted the prospecting expeditions that discovered the African Copper Belt and who had had a ship built and carried in parts across Africa until reassembled to provide the first commercial service on Lake Tanganyika (which inspired the film *The African Queen*).

On its maiden voyage, with tides and weather against it, it failed to reach the Berkeley River by nightfall, as planned, and had to make for a mangrove-lined beach to lie over. After its bottom had scraped on some unseen rocks the crew decided to drop anchor and wait till daybreak. Though concluding its voyage to the end of the navigable stretch of the Berkeley River, not quite as planned, the *Robert Williams* had shown it was completely unsuitable as a supply vessel. It was eventually towed to the Kalumburu mission station and given to the missionaries, while for supplies to Sally Port (the riverside camp) a chartered boat proved altogether more effective.

Given the Congolese connection of Tanganyika Concessions, it was only a matter of time before contact was made with the Belgian diamond-mining company Sibeka. Their Kinshasa-based general manager, a geologist named Bruno Morelli, visited the KJV follow-up operation in the northeast Kimberley. He made it clear that their strategy would have been to locate alluvial deposits, not necessarily large-scale, to generate cash flow while they mounted a kimberlite search. He clearly was no lover of kimberlite exploration.

As Chris recounts the story: 'Bruno took a gold pan, pick and shovel out to the King George River and spent some hours digging out crevices in the rapids and panning down the gravel. When we returned to pick him up, he proudly showed us the diamond he had found there. "Now you see *la méthode belge*! The diamond is a slippery fellow, like a fox. When you have the fox by the tail, never let him go! There are diamonds here. Why don't you dig them up?" From that moment on, we informally named this stretch of rapids on the King George River "Morelli's Fox".

'In our opinion there was very little gravel around, insufficient to support any sizeable diamond mine. Morelli said, "Ah, Smith, you never find the kimberlite; and if you find him he will have no diamonds." This, after all, had been Sibeka's experience in Tanganyika.

'As a result of Bruno's visit, the JV had no option but to test gravels around the northeast Kimberley for diamond content, although the kimberlite search continued in parallel.'

Soon after Morelli's visit it was clear that the operation needed a bigger plant to test large samples of gravel. A 5-tonne-per-hour Dense Media Separator (DMS), a distinctive and, for a geologist, readily identifiable piece of equipment, was parked in Wyndham while preparing to embark on a 1700-kilometre trip to the King George River area. It is probably no coincidence that there happened to be a De Beers geologist in Wyndham at the same time. He trailed the DMS for two weeks; this took him to Kalumburu. Chris Smith, camped nearby, got a message that the De Beers man wanted to come and talk to him. Not surprisingly he made it abundantly clear over the radio that visitors were not welcome, and the geologist left. The KJV's cover had been blown.

In February 1976, with the capacity of various joint venture partners for funding nearing exhaustion and no diamond mine in sight, they got their breakthrough.

◆◆◆◆◆

CONZINC RIO TINTO Australia, or CRA, had been approached by Ewen Tyler with a view to farming in to the joint venture. Their management, headed by MD Sir Rod Carnegie himself, visited the field operation and were impressed with what they saw. When they consulted with parent RTZ in London, it was suggested that they get some expert advice. The South African office had no hesitation in recommending Baxter Brown as one of the best diamond geologists anywhere, with a sound grasp of all aspects of exploration.

Baxter was bullish about the prospects of finding new world-class mines. In a well-tramped part of South Africa the Venetia pipe was discovered in the mid-1970s, after De Beers had been spread across the length and breadth of the country for nearly a hundred years. How much greater were the chances in a remote corner of Australia where helicopters had only recently replaced horse-back travel, eyes always high above the ground. Baxter had been impressed with the KJV laboratory in Perth and had pored over maps of results from the Kimberley and peered through his loupe at samples from there. As the room filled up and he prepared to give his presentation, he was surprised to see the legendary RTZ chairman, Sir Val Duncan, walk through the door. Baxter waxed eloquent and Sir Val liked what he heard. There was some discussion of the details and the decision was taken. Baxter was to be retained. The new association, constituted in February 1976, was named the Ashton Joint Venture (AJV), after the Ashton map sheet in the central Kimberley, where no suggestion of diamonds had been recorded. There was no point in making life easy for the competition, and competition there now was.

The Rio Tinto (formerly CRA) Australian head office in Melbourne.

Soon came the important discovery. In following up anomalous radioactivity near the Pteropus Creek discovery site (many kimberlites are weakly radioactive), Mick Paltridge and Chris Smith walked on to an unusual-looking rock outcrop which Chris suspected might be kimberlite-related. The laboratory confirmed his suspicion: Australia's first 'kimberlite' – in fact an olivine lamproite, subtly but importantly different from a true kimberlite as classically defined – had been discovered. A later sample yielded one small diamond. As if to prove Professor Prider's assertion that the lamproites were kimberlite-related, the first true kimberlite was found quite near Pteropus Creek, at a locality they had named Skerring. Bruno Morelli would have smiled when he heard it was completely sterile.

The CRA geologist assigned to the AJV, Frank Hughes, features larger and larger in

Chris's notes. He describes him as 'a highly skilled field observer and geological mapper who loved nothing more than being out in the bush'. When, towards the end of 1976, the joint venture had found indications of kimberlites far to the south on and near a farm called Ellendale, 'Frank Hughes insisted he should follow up this result as my borrowing his helicopter would just be a disturbance to his programme. In hindsight, I realise that this was an important moment. CRA had to make a decision on whether it should increase its equity in the JV and take over management. Frank always had a good instinct for what were important, key, priority issues. He took himself to the indicator cut-off point where there was a small stream gulley, found a nice shade tree, and proceeded to dig up the stream and recover diamonds by hand jigging. He also took loam samples from the surrounding countryside, and these were sent to Perth laboratory where they came back with widespread indicators and microdiamonds suggestive of a large pipe. CRA took the decision to increase its shareholding and become JV manager.'

Ashton Joint Venture samplers collecting wash from a dry riverbed to check for kimberlite indicator minerals.

The Ellendale area is flat, with little exposure, and exploration is difficult, but in 1977 sampling and geophysics located a 76-hectare diamondiferous olivine lamproite pipe, the third largest pipe then known anywhere in the world. Preliminary sampling suggested the grade was low and the diamonds small. Since the magnetic survey had shown the pipe clearly, it was decided to fly an airborne survey over 5500 square kilometres of the most prospective ground. A lot more pipes were found and a small plant established. Hopes were raised to a new level when Frank picked a 1.75-carat diamond off the surface.

As the surveying continued, a total of 46 lamproite pipes were found, several of them running to tens of hectares. And just as a protracted spell of finding diamonds and indicator minerals prior to the discovery of the first pipe had proved tantalising to the Kalumburu Joint Venture, so was the evaluation of the Ellendale field to the AJV. But though the stones were of superior quality, the grade was low to very low. When two of the pipes were bulk-sampled, the diamonds, while predominantly of gem or near-gem quality, proved small or very small and those from the smaller pipe were generally of poor colour, the value of the beautiful intense fancy yellows being appreciated only by later owners.

By 1979 Ellendale was looking more and more like a near-miss and most of the prospective area had been covered. As project managers, CRA decided the area should be dropped and the regional reconnaissance continued. Chris is full of praise for the CRA management.

'Without the ongoing support of top management, the AJV could have ended there and then in 1978 after the disappointments of Ellendale. But CRA believed in exploration and continued to back us. Ewen Tyler remained as chairman of the JV, always charmingly holding the partners together and encouraging us.'

◆◆◆◆◆

IN AUGUST 1979 one of the sampling crews was moving close to the eastern boundary of the Kimberley, near where some kimberlite dykes had been discovered in 1977. They were sampling creeks east of the Great Northern Highway between Hall's Creek and Kununurra, an arbitrarily chosen eastern boundary of the Kimberley craton.

Using helicopters to get into the valleys, Chris remembered during the initial sampling seeing a helicopter flying high overhead – but not out of sight – towards Hall's Creek. They had seen fuel drums alongside the road and recognised the trucks delivering them as belonging to De Beers. Later they would realise that the diamond giant's helicopter must have flown more than once right over the Argyle pipe.

The newlyweds Maureen and her geologist husband, John Towie, were assigned the task of sampling the area east of the Great Northern Highway. On 28 August 'our laboratory reported that three of their routine stream samples at 5-kilometre intervals taken from Smoke Creek, which drains into Lake Argyle, each returned diamonds and chromites. In a story reminiscent of Madame Sarsadskikh and the discovery of Zarnitsa in Siberia, Maureen was pregnant and had to return to Perth. John stayed in the field, and he and Frank Hughes sampled upstream of the discoveries. At first they thought the stones must be coming from Devonian conglomerates that outcropped in the ranges, but the train led them further and further upward into the headwaters. It was also apparent that the local gravel terraces contained high grades of diamonds and would probably be mineable. Warren Atkinson [CRA's exploration manager for Western Australia] flew up to the site and on 2 October 1979 he and Frank Hughes walked into the valley in the ranges at the very top of Smoke Creek and recognised outcrops of bedded volcanic breccia and tuff there. Thus was the Argyle lamproite discovered. Frank raced down to Ellendale with a piece of the tuff to show me; you could see likely olivine pseudomorphs in it.

'Now the pegging had to be done. Camp was kept at Kingston's Rest so as not to alert any potential competitors. Frank Hughes hired every rentable vehicle in the neighbouring town of Kununurra and had them parked in the CRA exploration office yard in the town. He also purchased every available topographic map of the area from the government survey offices around Australia. Then as many men as the JV could muster would fly out each day to the Argyle pipe and peg claims over and around it. So we thought the pipe and neighbouring alluvial deposits had been safely secured.'

Though they might have hobbled the competition in the field, there was still a problem, potentially far more serious than being spied on or outflanked. An Australian 'junior' company thought they had found a loophole in the timing of the AJV's reapplication for the licence covering the ground over Argyle and were prepared to go to court over it. Only the intervention of the Premier of Western Australia saved the day for the joint venture. He recognised that millions of dollars had been spent in getting the AJV to the point of discovery and that the apparent loophole was as much the fault of the state government as AJV's. Finally, on 11 December, CRA made a public announcement of the Lake Argyle Prospect.

◆◆◆◆◆

THE CRA-MANAGED Argyle Diamond Mines moved fast. Three years later, with the feasibility study of the alluvial and kimberlite deposits at an advanced stage and plans in place to bring the alluvial resource into production the following year (1983), a marketing deal was struck between CRA (56.8 per cent) and Ashton Mining (38.2 per cent), acting jointly, and the Central Selling Organisation (CSO). The minority shareholder Northern Mining, controlled by Alan Bond, would deal directly with an Antwerp diamantaire.

What enabled Ashton Mining to raise the substantial funding it would need to develop the mine was the guaranteed offtake of Argyle's production by the CSO, at a value which at the time seemed reasonable. Without any in-house valuers or staff to classify and prepare parcels of stones for valuation, there was no option but to go the CSO route, at least to start with.

Argyle diamonds in the rough.

Already the mine was slated to produce 20 to 25 million carats a year, making it the world's biggest producer, singlehandedly eclipsing production from De Beers' South African mines. Nevertheless, with results of large bulk samples to hand, it was clear that the value would be far less. The Argyle stones were small, with a high percentage showing poor colour, although the not uncommon fancy pinks would lift the value somewhat.

Though low-grade and tucked away in the aptly named Ragged Ranges, Argyle was well worth finding. Even the financiers were happy. Not only would the large, deep-reaching pipe give the mine in excess of 100 million tonnes of ore, at an exceptionally high grade (7 carats per tonne), but mine construction would start with the alluvials. This would offer low-cost mining and early production, in addition to which the gravels were higher-grade with bigger, better stones. The implications for cash-flow were enormous.

From the earliest days Argyle has been a success story. In terms of the carats produced, it has been the world's single biggest diamond mine for nearly 25 years and has transformed its owner, Rio Tinto, into a major player in global diamonds. As I write, in early 2010, the open pit is reaching the end of its economic life and plans to take the mine underground are on hold while the outlook for diamonds improves.

The mine's success is a tribute to its management. Since the first day the high grade has had to be balanced against a very low-value product, taking the production as a whole. In 1996 the contract with the CSO/DTC was terminated and since then Rio Tinto – the majority owner now that CRA has been absorbed into the parent company – has undertaken its own marketing. It doesn't take much imagination to realise that sorting and valuing over a hundred million tiny diamonds a year (hundreds of thousands a day), stones which hover on the borderline between industrial and near-gem quality, is a mammoth undertaking.

Between 30 000 and 40 000 carats a year of Argyle's production – around a tenth of a per cent – are fancy pinks, which rarely are dark enough to be called red. These are practically unique to Argyle and may sell for $100 000 per carat if they are bigger than 1 carat. Even rarer are blue and green diamonds, also with price tags at the top end of the scale.

As for Ellendale, it continues to tantalise. Since CRA let it go, various companies have tried to trim the sails and make the two pipes on which CRA concentrated pay, without a great deal of success. The latest is Gem Diamonds, which has turned the far lower-grade Letšeng pipe in Lesotho to good account, but only because of the astronomically high value of its beautiful diamonds, including the occasional giant. Though Ellendale stones are far better on average than Argyle's, they are not a patch on Letšeng's.

A deep pink Argyle diamond.

Lamproite: a new diamond-bearing rock type

A brief word on the geology. We started thousands of kilometres to the east of Argyle with alluvial stones in 'deep leads' in New South Wales. They were found far from any craton, with no kimberlites anywhere near, and to that extent remain one of the unsolved mysteries of Australasian diamonds.

Even when we came west, we saw that one of the leads that put Ewen Tyler on the trail was leucite lamproites, rocks which his professor had proposed were related to kimberlites, but with some fundamental differences. He knew that though they were the plugs that stood out as prominent features in the topography, they themselves were not a target.

In the first literature on the Western Australian diamond geology, the entire suite of rock-types – from true kimberlite at one end of the mineralogical and geochemical range to leucite lamproite at the other – were called 'kimberlitic rocks'. Neither of these 'end members' was payably diamond-bearing: the economic and near-economic pipes at Argyle and Ellendale were compositionally in the middle – olivine lamproites. Nowadays this is what they are called. The differences between the types are clearly evident to diamond geologists, which is what is important. They need not concern us here.

Also different is the structural setting of the Western Australian mine lamproites when one compares them with just about every other diamond field around the world. Whereas in Africa, Siberia and Canada the diamondiferous kimberlites are commonly well in from the margins of the cratons, in the Kimberley they are right on the edge. The different geology had geologists scratching their heads for a while, then the model was realistically revised and everyone was happy.

AUSTRALIA IS A COUNTRY where junior companies abound and thrive, as in Canada. Many of them ply their trade in diamond exploration, having done so for over 30 years. Major companies like De Beers, Rio Tinto and BHP Billiton have honed their exploration skills there. Yet Argyle, discovered in 1979, was the last important diamond discovery. For nearly 140 years, while the source of the river diamonds was identified for the first time in the Cape and Dr Tom Clifford made the pivotal connection between diamonds and cratons, Australia continued to throw up more questions than answers.

It vexes geologists and has tested explorationists and miners. But thanks to a large number

Aerial shot of the Ellendale camp.

of doughty individuals, technology and persistence have triumphed. Two joint ventures, then CRA and finally Rio Tinto, all adapted to a game with new rules: lamproite was found in the field, not kimberlite; chromite, not garnet, in the laboratory; and, lastly, small discoloured diamonds, with champagne-coloured stones rare enough to have corks popping, instead of thumbnail-size 'blue-whites'. They have run off the field of play as winners. For years the Ashton mine produced a third of the world's diamonds from one fabulous but elusive pipe. It is an inspiring tale.

CHAPTER ELEVEN

MODERN EXPLORATION AND MINING

More people mine diamonds for their own account than any other mineral on Earth. And every diamond miner, before he got his first mine up and running, started as a prospector. Where does prospecting end and mining begin? The answer is that often there is no clear break: one merges imperceptibly into the other. This could be said of just about any kind of artisanal working, whether for tin in Malaysia, cobalt in the Congo or diamonds in Mato Grosso. Where diamonds are different is that the gradual elision from exploration to mining may happen just the same in multimillion-dollar companies, particularly when they are mining alluvial deposits.

Every other kind of major mining venture, including the big kimberlite-hosted diamond mines, goes through a progression of stages with each phase ending at a strategic decision point. The last two of these are reached when the pre-feasibility study (PFS) and the so-called

De Beers' Jwaneng Mine in Botswana exemplifies what major companies hope to find with their multimillion-dollar budgets.

bankable feasibility study (BFS) are presented to the board of directors. With a positive BFS the necessary millions, or billions, of dollars can be raised to build the mine. Until the BFS is submitted and approved, every activity that has taken place forms part of exploration: the distinction between exploration and mining is absolutely clear-cut.

The exploration needed to get to a BFS will generally have cost millions – money into a hole in the ground, say grizzled financial directors; many holes in the ground more likely. In a copper prospect, for example, good geology and low-cost soil sampling may give the first indication of a potential deposit below the surface. This may lead to trenching, still quite inexpensive, which may confirm the existence of copper mineralisation of a kind that could make a mine. Gradually the exploration gets more and more costly, as geologists look deeper into the ground and the likelihood increases of what has been discovered being commercially viable. Years go by and millions of dollars are spent before the day arrives of which everybody on the project has dreamt. The BFS is declared positive. One day, in a few years, a mine will be built: what started as bare veld or pristine jungle or untouched tundra will see a giant plant standing and a village, and before long – out of sight – a terraced hole in the ground with monster trucks trundling in and out.

Around the world equipment like this moves many millions of tonnes a year.

As we've said, diamonds are different. Firstly, with a lot of luck and careful planning from the outset, the diamonds recovered during exploration might ensure that this phase will start to pay for itself soon after inception. Secondly, a huge amount of material will have been tested, many tens of thousands or even hundreds of thousands of tonnes, by the time a decision is made to mine the deposit.

To the casual observer who drives past the diamond operation from time to time, the transition from exploration to mining is not obvious at all. He hasn't noticed that the plant established for the testing of the deposit has been enlarged or that the fleet of excavators and trucks is bigger. Nor would he know that the economies of scale have moved the operation from break-even to profit.

The reason why diamond mining is so different from this is best answered by a term used by geostatisticians: the nugget effect. Take a block of rock, or river gravel, the size of a single

Gem Diamonds' Cempaka prospect in Kalimantan, an example of a small trial mining operation.

bedroom, and imagine that somewhere within it lies a diamond the size of a raspberry or a gold nugget as big as an apple. That it would be worth your while to process all the material to recover the diamond or the nugget is beyond doubt. Equally certain is that you would have to process it all, because you've been given no clue as to where the hidden treasure is located and, even when you've found one, whether it was the only one. Raspberry-sized diamonds are rare. A few pea-sized, top-quality stones might be worth as much. The only way of knowing whether they're there or not is by carefully processing every last kilogram of that hundred tonnes (the amount of rock or gravel that would fit into a smallish bedroom).

So to achieve the same level of confidence that the copper prospector has about his deposit, the diamond miner has to test a huge volume of rock or gravel. Think of it like this: firstly, diamonds are worth more than a million times the value of copper, and, secondly, they occur as discrete crystals, often quite large, here and there, not as finely divided mineral spread evenly through the ore, like copper or gold.

Another aspect of the large sample size is that to an appreciable extent the principle of economy of scale is as relevant to exploring for diamonds as it is to mining them. The more material you process in prospecting, the closer you get to breaking even or even, in exceptional cases, showing a profit. This certainly never happens in any other kind of company exploration (nor, by the way, in a De Beers or Rio Tinto or BHP Billiton diamond mine, where the exploration costs a lot more for a variety of reasons).

'Suck it and see' is a basic philosophy that more than a few junior and middle-size companies apply to the exploration and mining of diamonds. They are gamblers, and if they've done their homework properly – and that's a big 'if' – the odds are stacked in their favour. The potential reward, they know, compared with the outlay, is enormous. De Beers, on the other hand, evaluates a big diamond-bearing kimberlite pipe very much as Rio Tinto would a potentially world-class copper deposit. It will take years and years of very professional, high-tech input before a BFS is presented, and the diamonds recovered are of almost inconsequential value compared with the cost of exploration. The outside observer would have no doubt at all when the progression had been made from evaluation to exploitation.

LET US LOOK at the very basic question of how people find new deposits of diamonds. Before we look at some illustrative examples, we should review one general principle. 'If you want to hunt elephants go to elephant country,' is a dictum as relevant to the prospector with his mule, shovel and sieves as it is familiar in the corridors of a Top 100 company. The prospector in the outback of Mato Grosso will look to see where the metaphorical elephants are being shot and get himself there, determined to find the survivors. The exploration manager in his air-conditioned office in London or New York will study the world map on his wall: he will search for patches of jungle or veld so remote and inhospitable that the elephants still roam. In his inner sanctum, where his secret is safe, his map shows the latest delineation of Earth's cratons. Deep under those cratons, conditions of temperature and pressure in the mantle combine for the hallowed 'diamond stability fields' to exist. Beyond those special preserves, we are told, all carbon in the mantle is worthless graphite. Research by universities and geological surveys is constantly updating and refining the demarcation of cratonic crust. Sometimes opportunistic discovery of a new diamondiferous kimberlite may itself confound the cognoscenti and bring them either to redefine the edge of the craton or modify the model.

Our exploration manager's map is like no other: his researchers update it monthly, using data dug from deep in scientific journals. Every known kimberlite pipe and dyke is marked on the map with a coloured pin, sterile in blue and fertile (even if subeconomic at today's prices) in red; alluvial diggings are shown with green pins. Every month new pins are added. He sees a gap: it's on the craton, green pins around it, on rivers draining the craton. There is no time to waste. He must prepare for his safari. What are the politics? How is the mining law? Is there any infrastructure? Will he need helicopters? What are his own budgetary constraints? He addresses these issues, and a thrilling adventure begins for him and his team.

As a first example of the metamorphosis from concept to prospect to mine, let us home in on the Orange River where it winds through the Richtersveld, some of the wildest country anywhere. We're starting here because the Oena Mine we're going to be looking at exemplifies a

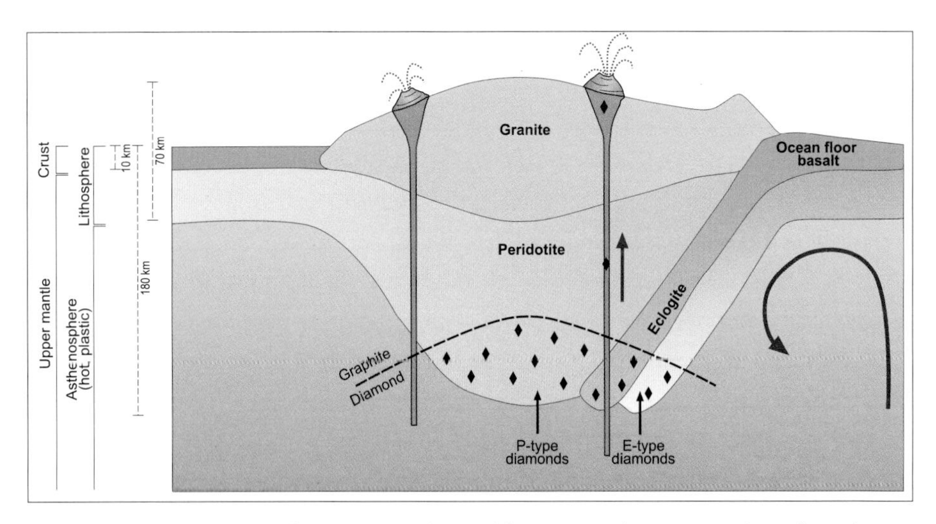

A figure showing how geologists imagine diamond formation and transport to the surface. The peridotite has formed part of Earth's mantle since very early times: the eclogite, by contrast, is forming continuously as oceanic crust and mantle are subducted under the continents.

project where intuition backed by common sense found and turned to account a pocket of gems of unsurpassed richness. It was a discovery in which university graduates were conspicuous by their absence. Yet as a mine it drew more than one grey-haired professor of geology, to visit and revisit, such was its magnificence as a demonstration of how rivers concentrate diamonds.

The legacy of Oena is a tribute to one man, Dick Barker. Dick is himself part of a diamond dynasty. Near the beginning of George Beet's classic book *The Grand Old Days of the Diamond Fields* there is a picture of Dick's great-grandfather, with the caption 'Captain Roderick Barker, the Grand Old Man of the River'. Beet describes him as 'the hero of many untoward incidents in Diamond Fields territory.' He groups Roderick Barker with three other acquaintances of his, including Cecil Rhodes, all leading figures of those days, 'any one of [whom] would have succeeded in directing wisely the destinies of a continent'.

Dick's grandfather, his father and his two sons, Kevin and Maurice, have all been deeply immersed in the diamond business, but it is Dick who is best known to hardened diggers and leading professors across South Africa. In the late 1980s Dick heard of a riverbank concession in one of the wildest and most inaccessible parts of South Africa, the Richtersveld, where there were alluvial terraces similar to those being mined lower down the Orange River at the fabulous Octha Mine. Knowing that the rights-holder was not exploring the terrace and showed no signs of doing so, Dick resolved to secure the concession for himself. It was state land and the principle of 'use it or lose it' applied. He shaped his strategy with exemplary thoroughness, to the point that the State President had no option but to terminate the previous licence-holders' rights. To their chagrin and after two years of waiting, Dick's application for prospecting rights was successful.

To get to the concession, Barker and his small team had to traverse 28 kilometres of rocky cart-track, accessible only to the most robust 4x4 vehicles. In the end they got a front-end loader, an old Ford truck, his cylindrical sizing screen and a six-foot rotary pan into the area of terraces. In the first three months they recovered four diamonds with a total weight of a carat from the so-called Meso Terrace, not far above the banks of the river.

Dick decided to try the patches of the higher Proto Terrace that he had seen. There were three of them in a line parallel to the river: he would start on the middle one. Before he made the long trip to Kimberley to sort out his finances, Dick gave his foreman instructions on where to start excavating what he had decided to call T1. In the Griqualand West capital he found his bank was about to foreclose on him, preparing to seize his farms and his house. In a fury he changed banks, securing one with more belief in his project.

His rage returned when he got back to the project and found that the foreman had chosen another part of T1 to start the excavation, though he was somewhat mollified when he saw a 6-carat perfect white diamond come out of the pan concentrate. Nonetheless, he moved the loader to the downstream end of the patch of gravel. On the fourth day – and here there is a catch in Dick's voice as he remembers it – they saw a number of diamonds coming out with the smaller-size fraction of the concentrate, stones of 3 carats, 7 carats and so on. Minutes later, bouncing and rolling out of the end of the inclined sizing screen tumbled a 20-carat, flawless 'D'-colour diamond. It was the first of many.

Dick had concrete blocks cast, which he put on either side of the strip of gravel, and between them he strung a length of fishing line. Measuring along the line and from the beacons with a tape, he made a crude survey of the digging, marking key points in it with paint. Later it would be accurately surveyed with instruments. As they excavated, every stone was registered once it had been weighed, and its exact provenance noted. Six years – and 33 000 carats – later, the Barkers could believe it when they were told that they had kept 'the most complete records in the world'.

They were beautiful stones, every one of them, and at an average size of 1.8 carats they sold for a higher value per carat than from any other mine in the world. The biggest stone recovered was 78 carats, not so big, to be sure, as further up the river where in the heyday of the deep potholes and riffles along the Vaal River around Barkly West, stones of over 100 carats were, if not a regular occurrence, at least more than one or two. Even so, at today's price of, say, $25 000 per carat, a 78-carat flawless 'D'-colour diamond would be worth nearly $2 million.

But it is not for the dollars he made at Oena that Dick Barker will be remembered; it is for his determination to see what treasures his T1, T2 and T3 terraces held and, even more, for the records he kept. Having got to know Dick a little, I believe I am correct in saying that his record-keeping was carried out not only, perhaps even not mainly, to know where to find more diamonds, but just so he could understand why they were where they were. It wasn't

Dick Barker, self-taught alluvial sedimentologist extraordinaire.

just the diamonds that he kept track of: it was the bedrock topography, and the nature of the gravel, and where the bigger-than-5-carat stones were concentrated, and other factors besides. These were all, he knew, integral parts of a detective story as gripping as any. A big part of Dick's quest is understanding for its own sake.

That is why, when it comes to comprehending sedimentary processes and their influence on diamond concentration, Dick can hold his own with any university professor or the most respected exploration manager. This is another of the multitude of ways in which diamonds are different. Alluvial diamonds are a great leveller. In no other field in mining can practical people, without important-looking letters after their names but with sharp eyes and minds that never rest, hold the wisest intellectuals in their thrall as they offer their perception of geological processes and phenomena.

YOU MIGHT SAY that finding deposits of alluvial diamonds is as simple as following rivers that drain diamond-bearing headwaters; and for the most part this would be true. One of the most fabulous deposits of alluvial diamonds along that true giant of diamond rivers, the Vaal, was discovered at a place called Holpan, 40 kilometres north of Kimberley. It is out of sight of the Vaal, but only just, and its primary gravels and surface enrichments are clearly similar to those close to the river, just a little older. But (and here is where the rule of following the modern rivers breaks down) some alluvial diamond deposits are so far from any river that for a long time they defied explanation. In fact their very existence triggered research that has led to an entirely unexpected model of river evolution and, consequently, landscape formation.

The best example of alluvial diamonds along an ancient palaeo-river that has disappeared without any obvious trace is found in a few small workings, now defunct, around the village of Brandvlei, in the Northern Cape. It is just about the most arid desolate country you can imagine, liberally strewn with vast pans or playas, hundreds or even thousands of hectares in extent. They are hard, flat and so level that one of them, Verneukpan, was chosen by Malcolm Campbell for his assault on the world land-speed record in his Napier-Arrol-Aster Bluebird in 1929.

The countryside is so dry that farmers only dream of swollen tree-lined rivers, yet aeons ago – between 70 and 30 million years – this is where the great Karoo River crossed the South African hinterland to the Atlantic coastline. It drained what we know as the Orange and Vaal river basins in the central highlands, stripping off the tops of young kimberlite pipes and releasing their diamonds to start their long journey westwards. Most of the stones made it to the coastal plains to form the oldest deposits there, but a few were caught up on the way, like those around Brandvlei. For now it suffices to say that not all alluvial diamonds are close to the rivers we know. For the most part, however, finding alluvial deposits means staying close to rivers draining diamondiferous headwaters and applying the few basic principles that Dick Barker did.

Verneukpan, one of the huge salt pans, or salars, common in inland Namaqualand.

On the other hand, it is in finding diamondiferous kimberlites that we move a short way towards rocket science. Yet the process hasn't always been as high-tech as it is now. Near the dawn of systematic diamond exploration the distinctive mineral composition of kimberlite rock gave prospectors a technique that could very effectively be turned to account in locating pipes. Some of the minerals that occur in kimberlite are practically unknown in other common rocks. These are a garnet called pyrope, which although it looks like the common garnet, almandine, is chemically quite different; and chrome diopside, the beautiful bright green mineral that is extremely rare except in some kimberlites. Another of the so-called indicator minerals – or KIMs (kimberlite indicator minerals) – is ilmenite, a black, weakly magnetic iron and titanium oxide. Unlike the other two, ilmenite is a comparatively common mineral found in various rock-types, but the combination of the three – pyrope, chrome diopside and ilmenite – in a soil or river-sand sample is what widens the diamond prospector's eyes and sets his pulse racing.

You wouldn't find them together in a sample at all were it not for the fact that in temperate or cool climates they don't break down easily, particularly garnet and ilmenite. Though they may be relatively infrequent trace minerals in kimberlite, they are a hundred times more common than diamonds. Their durability means they travel just about as far, and they're all heavy – diamond and pyrope garnet have practically identical specific gravities – so they concentrate in the same sort of places as diamonds do, faithful travelling companions.

All the same, these minerals are just as common in sterile kimberlites as they are in the ones explorers want so badly to find. High technology and intense research would be required to answer the question whether a pyrope is a pyrope is a pyrope, or whether there are subtle differences that could be used in exploration. Decades passed between the recognition of KIMs as a valuable prospecting tool and the realisation that the pyropes from fertile kimberlites were chemically ever so slightly different from the others.

Zeppelins or blimps or dirigibles are suitable for some forms of airborne geophysical surveys, being less susceptible to turbulence than conventional aircraft.

The branch of earth science we call geophysics exploits the fact that rock-types differ from one another in their physical make-up: their density, porosity, radioactivity, magnetism and electromagnetism. This difference can be detected by instruments that measure these particular properties. The differences may be quite substantial or they may be so slight that only the most sophisticated equipment will show them. Instrument manufacturers never stop looking for improvements that will make their equipment more sensitive, more versatile and more affordable.

In the 1960s, when Chris Jennings started in his quest to apply geophysics to finding new water reserves in that driest of dry countries, Botswana, his tools were basic. But they worked, and by their judicious application and refinement he was soon discovering important new resources deep in the Kalahari, that huge deposit of savannah-decked sand that makes up most of Botswana, then a sparsely populated country relying on ranching and ecotourism for its revenue. Some of the boreholes tapping the water he found were deep, to be sure, but huge new areas were opened up to farming and human habitation.

Jennings's vision extended way beyond water. He knew, as did every explorationist, that new mineral deposits were being discovered by geophysics. If water was his commission, diamonds were his passion. Part of Jennings's function, as deputy director of the Botswana Geological Survey, was to liaise with the companies that had been given concessions to explore for minerals by the Mines Department. It was inevitable, then, that he knew of De Beers' discovery of a group of kimberlite pipes in the east of the country very soon after it was made in 1969. It was a ground-breaking find which in a couple of decades would turn Botswana from one of a long list of the unendowed into the world's biggest producer of gem diamonds. Exploration continued and before long word was out that De Beers was developing a major new mine, to be called Orapa. Suddenly Botswana was elephant country.

At this stage Jennings moved to Johannesburg to head the exploration division for the Canadian miner Falconbridge. It was not long before an airborne magnetometer survey had vindicated Jennings's belief that geophysics would penetrate the thick Kalahari sand to locate kimberlites. Having kept the news under wraps for a long time, in the early 1980s Falconbridge announced the discovery of a new field in central Botswana. The pipes had been tested by drilling, and the sampling of one of them, called Gope, suggested a payable diamond grade. It slipped, however, from Falconbridge's grasp, and to this day has not been turned to account, mainly because of the quantity of sand to be moved before the surface of the pipe would be reached. Technically it was an exploration success story.

Jennings was then transferred to Falconbridge's head office in Toronto. There he found himself at an early stage of what turned out to be a long and frustrating learning curve. He would discover just how difficult it was to infect grey-haired, grey-suited accountants with the enthusiasm and endurance needed to follow the faint but unmistakable trail that would lead to the biggest prize of all: the fabulously rich kimberlite pipes of northern Canada. While he had been thwarted in Botswana, it was not half as much as he would be in Canada. In the end his persistence would be rewarded, but only once he had abandoned the major mining companies, with their leviathan slowness, and linked himself to more mobile 'juniors', before founding his own company, SouthernEra Resources, early in 1992.

Chris Jennings and his SouthernEra would go from strength to strength. But that is not the point. The reason he's mentioned here is because of the pioneering work he did in the application of geophysics to exploration for kimberlites. Nowadays no serious kimberlite explorer would dream of excluding geophysics from his arsenal of primary target generation. Why does it work?

A kimberlite pipe that comes punching up through the crust to reach the surface is invariably different in at least one physical respect from the rocks through which it cuts. The difference will depend on two things: the kimberlite itself and the surrounding rock. The measurable contrast may be in any of the properties mentioned previously. The kimberlite may be more magnetic (ilmenite, a common kimberlite component, is magnetic) than the host rock; it may be denser; it may be more porous, leading to a breakdown of some of its component minerals to clay, which conducts electricity; or it may be marginally more or less radioactive.

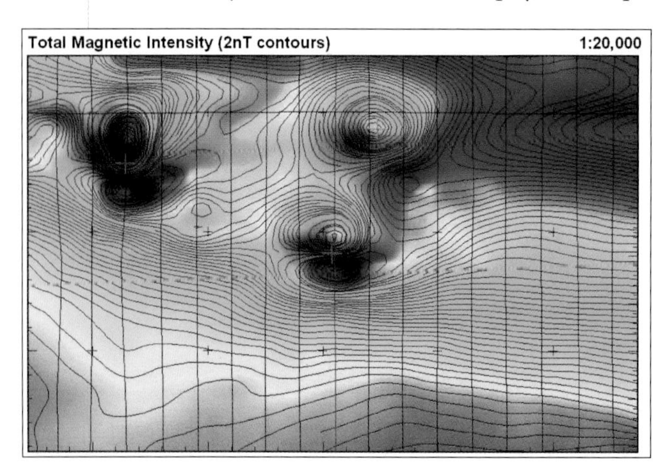

An 'aeromag' (airborne magnetic) survey map, with flight lines 250 metres apart, shows potential kimberlite where 'highs' (red) and 'lows' (dark blue) are next to each other. This is called a 'bipolar' anomaly.

Chris Jennings was well enough versed in all the nuances of geophysics, sound enough in his understanding of geology and resolute enough in his quest for diamonds to have an edge over his competitors. But he had also a new secret weapon that would outrank any of those: geochemistry, and this was a weapon he would not have had without the enormously powerful ally of John Gurney, Professor of Geochemistry at the University of Cape Town (UCT). You cannot open a book on the recent history of diamond exploration without finding the name John Gurney liberally dotted through its pages, whether in the chapters on Canada, Botswana, South Africa, Lesotho or Namibia. Gurney has been there, and he has seen dozens, if not hundreds, of rock samples from scores of kimberlite pipes and dykes in all of these places. He doesn't work alone, though, nor is UCT his only stamping ground. His commercial laboratory at Mineral Services – paired with the university facility – is at the forefront of geochemical research into kimberlites and their minerals and enjoys unqualified global prestige.

In arguably the most important scientific breakthrough in the history of diamond exploration, John and his team found that pyrope garnets in 'fertile' kimberlite have a chemical make-up that differs from those in sterile pipes. It is a very subtle difference and there are exceptions, but as a general rule it applies.

The work of Gurney, the Russian Nikolai Sobolev and others on kimberlite geochemistry, though crucial to its diamond exploration application, is not restricted to that realm. Just as important in its way is the information his research has yielded on temperatures and pressures in and beyond the 'diamond stability field', which have direct bearing on the structure of the different layers in the mantle. Together with the geological work of Tom Clifford on cratons, Gurney's geochemical research has enabled us to unravel the mysteries of the character of the upper mantle and the crust above it. Clifford was the first to recognise the craton–diamond linkage, giving rise to what has become known as Clifford's rule, but he did it empirically, with no attempt at analysis. It is Gurney and his colleagues who have explained the relationship, using pyrope garnets and other minerals that kimberlites have scavenged from the mantle on their way to the surface.

◆◆◆◆◆

WITH ACCESS TO the substantial armoury of technology that we have reviewed, how does a new explorer begin? Firstly, let's imagine he has decided that diamonds have a future, and he has found an unexplored country where potential has been established. He has reason to believe it is elephant country. Where does he begin?

Drilling of an alluvial target to establish the presence and thickness of potentially diamondiferous gravel.

Perhaps some artisanal diamond digging was carried out years ago and the country has recently emerged from a period of political turmoil. The old diggings are his elephants. He knows there will be more diamonds up- and downstream from them and he knows they all came out of the jungle that fringes the river where the old workings can still be seen. He can choose to stick to the rivers and hope to set up a big alluvial mine, or he can decide to go after the kimberlite pipes – the big jumbos – in the jungle, or he can do both, depending on his funding. Let us assume funds are substantial and he goes for broke.

His alluvial team works up and down from the historical diggings, drilling through the shallow surface layers to locate the deeper boulder gravels that will carry any river-borne diamonds there are. Should he find substantial deposits of good coarse gravel near the surface, he will open them up and sample them, processing the samples through the plant he has set up. If he's lucky he'll recover diamonds and the grades will pass muster. He delineates his resource with further drilling and continues with strategic bulk-sampling. With 20 million

BAXTER BROWN

From the rainforest of Venezuela to the outback of Western Australia; from the gorges of the rugged Richtersveld to the wide horizons of the Kalahari, Baxter Brown is the diamond geologist *par excellence*. His passionate conviction that the Kimberley of Western Australia must be fertile terrain for diamonds helped persuade Rio Tinto to explore it and find a mine that would supply a third of the world's diamonds for decades. He recognised the strange-looking rocks releasing diamonds into the drainages at Guaniamo in Venezuela as kimberlite long after he had unearthed fabulous diamondiferous river gravels along the remote lower reaches of the Orange River. Sit with Baxter in his apartment, hear of his travels and explorations: incredulity sets in. The art on the walls, the titles in the bookcase, the music collection – all tell of a great connoisseur. Did he really do all those things, travel to those places? Go to his office; open his diaries. He was there. He has made good use of his seventy-plus years.

tonnes of gravel grading 0.75 carats per hundred tonnes, our explorer proceeds to mine the gravels. He has graduated: he is now a miner.

In the meantime his kimberlite team has contracted a geophysical survey company to fly a magnetometer survey over the area that his geologists have defined as being underlain by prospective craton. The survey maps show clusters of magnetic 'anomalies', which might be caused by kimberlite pipes. He commissions a contractor to carry out a gravity survey over the clusters to see whether any of his magnetic anomalies have gravity anomalies superimposed on them. Any that do are his priority targets and he does intensive ground magnetic and gravity surveys to pinpoint where he is going to drill.

One of his priority targets produces chips of kimberlite, blasted out from the depths by the compressed air that powers his drill and blows the chippings and fine dust out of the hole. He drills a grid of holes on a spacing of 50 metres by 50 metres, then closer, to outline the kimberlite pipe. It turns out to be roughly circular and about 300 metres in diameter, giving the pipe a surface expression of around 7 hectares. By world standards it's an average-sized pipe, certainly big enough to mine. Our explorer knows that most pipes are carrot-shaped, and so will reduce in size with depth. How much, though? He brings in a core drill to get some solid samples of the kimberlite, and by drilling inclined holes, at say 60° from the horizontal, he starts to get an idea of the underground shape of his pipe. He estimates from his core drilling and profiling that he has got approximately 13 million tonnes of kimberlite to a depth of 100 metres.

But is it rich enough to mine? What is its grade? He digs four pits across the pipe, spaced to give the most representative coverage, and takes 200 kilograms of fresh kimberlite rock from each of them, which he sends to a laboratory in South Africa. The lab does mineralogic and geochemical analyses of each sample, reporting that all four carry a high proportion of G10 garnets, and that the geochemistry of the other mantle-derived minerals, like ilmenite and chrome diopside, is favourable.

Buzzing with excitement, our explorer considers taking big samples with a Bauer drill, a huge rig which will drill holes of 2.5-metre diameter, for an estimate of his diamond grade. But because it is too expensive to bring one in from out of the country, he will use local earth-moving equipment on contract. First he must set up his plant to crush and process his 10 000-tonne samples. He is getting massive financial support from the government, which is keen to see a diamond industry developed, and he establishes a small plant.

His first sample gives him a grade of 25 carats per hundred tonnes (cpht), the second grades 15 cpht, the third 40 cpht and the fourth 10 cpht. The samples are all approximately equal in size, so the average grade for the pipe is 22.5 cpht. He's sold the stones on tender in Antwerp, for an average price of $120 per carat, quite heavily influenced by stones of 32, 20 and 15 carats. He has seen that most stones are less than a carat, with a sprinkling of stones between 1 and 5 carats, and the colour and clarity good but not exceptional. The price he has been paid is fair.

Drilling a kimberlite pipe in Botswana.

On the back of an envelope, while his phone call is put through to the consulting engineers who will compile his bankable feasibility study, he works out the value of the rock to 100 metres. It is $13 000 000 (amount of ore in tonnes) x 22.5/100 (its grade in carats per tonne) x 120 (the value of the diamonds in US$ per carat), or $351 million. He knows that to establish his mine and operate it for the length of time that it takes to mine to

100 metres will cost a fraction of that. Two months later his positive BFS is on his desk. He scales his plant up and buys his earthmoving equipment. He's ready to mine his first kimberlite pipe.

Would that it were so easy! All the same, even without experience in diamond exploration and mining you can imagine that what we've covered in a few short paragraphs takes many years and there will be pitfalls every step of the way. It will be an open-cast mine. The pipe is big, and to begin with there will not be much stripping of sterile country rock or host rock – into which the pipe was intruded – to be able to get at the kimberlite. As it deepens, more and more sterile material will have to be stripped away so that the slope from the surface down to the bottom of the pit remains safe. The 'waste' is dumped out of the way, to be returned to the

A typical bulk-sample testing plant, this one in Kalimantan.

pit when the mine is worked out, together with the 'tailings', which are what comes out of the plant once the diamonds have been extracted. All quantities and costs have been included in the BFS, so there should be no nasty surprises.

The plant is geared to process 150 tonnes per hour, so that the mine reaches the 100 metres depth after 10 years, having long since paid for its capitalisation. By now the grade has dropped to 20 cpht, but the stone price has increased to $150 per carat. Though the ore is worth more than when the mine started, the surface area of the pipe – at 100 metres below ground surface – is only 4.5 hectares.

Two BFSs are now carried out, one to analyse the feasibility of carrying on with the open pit, the other to investigate the commercial viability of going underground. The results show that the underground mining, even with the cost of a shaft and the other necessary engineering, is the way to go and will give the mine another five years of life. By the end of the mine's life 20 million tonnes of ore will have been mined, producing $600 million worth of diamonds at a very respectable profit margin. Some elephant.

Other than this second-stage deepening of what started as open mines, the only underground mining of kimberlite for diamonds is carried out on dykes, still called – strictly speaking, not correctly from a technical point of view – 'fissures' in South Africa. Moreover, it is practically only in South Africa that fissures are an important type of kimberlite intrusion and where the expertise has been developed over nearly a hundred years to mine notoriously thin ore bodies whose high grade more than makes up for their difficult mining.

For by far the most part, then, whether mining gravel in Brazil, kimberlite in Botswana or lamproite in Australia, diamond mines mostly consist of open pits, some of them enormous, others quite small. If you were to visit one of these you might notice a headgear close to the edge of the pit: this is because many of the older pits have underground mines below them, as in the hypothetical case we've just looked at.

THERE IS ONE last type of diamond mining that we need to talk about. The reason it comes at the end of the chapter is that, although it falls within the ambit of mining, it boasts no mines, at least not in any conventional sense. Stand on the beach and look out to sea: you may, if you're on a very specific stretch of coast in the southwestern corner of Africa, see a very strange vessel or two, but that's all you'll see of offshore diamond mining.

With the only important coastal alluvial diamond mining anywhere in the world happening along the Atlantic coastline in the Northern Cape and southern Namibia, it is no surprise that this is where the offshore activity is. The mining of submarine diamonds north of the Orange River mouth is an important part of De Beers production. It is a far cry from the close inshore diving operations by fortune-hunting daredevils who risk all to launch

A Debmar vessel used for mining the sea-bed gravel for its diamonds.

their puny skiffs into the raging surf of one of the most treacherous stretches of coast anywhere. The deep-water mining by De Beers would not be attempted by lesser companies without the vested interest. Many have tried: none has succeeded.

Dredging – another name for underwater mining – of a range of minerals has taken place for nearly 450 years, both in rivers and offshore. Nearly all of it has been in shallow water, for the obvious reason that costs escalate dramatically as the depth of mining increases. Where a single rig can be set up and maintained in a fixed position for years, such as in the exploitation of offshore oil reservoirs, profitable – even lucrative – operations abound. Once again, diamond mining is different.

We now know that there are huge reserves of diamonds on the bed of the Atlantic Ocean off the mouth of the Orange River, mostly to the north. The field extends from the wave zone to a depth of 120 to 140 metres and, like the onshore deposits, consists of beach gravels made up of boulders, cobbles, pebbles and sand. Diamonds and other heavy minerals like garnets are concentrated in trap sites in gulleys and potholes, just as they are on land, close to or on the bedrock. Difficult though it may be to imagine, the deposits were formed along ancient beaches when sea levels were correspondingly lower than they are today.

In some of the harshest conditions imaginable, with howling winds the norm and the icy-cold sea frequently whipped up into a foaming, raging maelstrom, the advances in geological understanding of what lies in the murky gloom below, and how it might be evaluated and mined, have been slow and costly.

In 1982, after twenty years of small-scale mining off the Namibian coast, first by a Texan entrepreneur, Sammy Collins, and then CDM Marine, seismic profiling and sampling had established that there was a world-class reserve of diamonds offshore, with unknown grade. De Beers Marine was formed in 1983 and continued research into the geology of the targets and optimum sampling and mining technology until 1988, when the project moved into the mining phase.

Geological exploration proceeded with seismic geophysics, visual observations from remotely operated underwater vehicles, and 'grab' sampling to pick up material for closer inspection, including boulders up to a metre in diameter. Sampling for grade evaluation purposes was conducted by small barrel drilling and bouncing digging heads.

After initial results using the original system of mining – with a digging head suspended from the ship – cast doubt on whether all the diamond-bearing gravel was being recovered, the first crawler system was developed until it was deemed to be 'cleaning' the bedrock effectively, leaving a minimum of gravel unrecovered. The crawler supplied material to a specially adapted 77-metre mother vessel equipped with the plant to produce a concentrate that could be helicoptered ashore for final sorting.

The Jago *is a two-man observation submarine which Debmar hires for a month at a time, giving geologists and engineers the unique opportunity to see and video-film the bedrock topography and the nature of the gravel.*

In the constant striving for improved recovery, experimentation with various systems led to adoption of the large diameter drilling system as the premier mining system, which would persist with many upgrades until the present. Two ships were added in 1991, both using this large diameter drill system, and the economies of scale thus achieved made it possible to mine grades that previously had been subeconomic. In constant pursuit of a more cost-effective operation, more vessels were added. The original crawler technology was also superseded by larger and more powerful machines and even larger support vessels. The fleet of production ships now stands at six, one of which operates off the South African coast, the rest in Namibian waters. Servicing both Namibian and South African ventures are company-owned sampling ships and various survey devices operated from contracted vessels. Despite their top-heavy and cluttered appearance, the ships are the pride and joy of their engineer creators, every carefully designed and positioned item on board serving its purpose. More importantly, they are capable of an annual production of a million carats of top-quality gem diamonds. Considering the conditions they work in, this is a remarkable achievement.

An artist's impression of a kimberlite volcano in its final stage of eruption.

CHAPTER TWELVE

THE FORMATION OF DIAMONDS AND THEIR EARLY TRAVELS

Diamond – the simplest mineral imaginable, yet with a history so convoluted that it has taken a hundred years to come close to understanding it. The key to solving the final mystery was not so much the slow unravelling that time brings as the arrival of a technology undreamt of a century ago, one which turned ordinary laboratories into a whole new field of exploration and discovery.

AFTER THE MOTHER lode of diamonds had been found in 1871 in what would become Kimberley, the next geological revelation that would shake the industry was made by Dr Tom Clifford in 1966, when he proposed an essential connection between diamond-bearing kimberlites and the ancient continental nuclei. Until then, no one had identified the true source of diamonds and it was only when an unknown blue-grey rock appeared in the early 'dry diggings' of the Cape that the scientific quest for the origins of diamond began in earnest.

For decades after the new rock had been named, it was assumed that the diamonds had formed in the kimberlite. The pipes, it was argued – correctly – were the solidified chimneys of explosive volcanoes, with the upper parts long since removed by erosion. An extreme example of explosive volcanism was provided by the spectacular annihilation in 1883 of the island of Krakatoa in the Sunda Straits between Java and Sumatra. There were also well-studied cases of fragmental flows on the slopes of volcanoes around the Mediterranean and in the volcanic mountain chains that rim the Pacific. It was logical to suppose that kimberlites belonged to the same genre of phenomena.

As for the diamonds, volcanic rocks with phenocrysts were common enough around the world and through the ages. These crystals, geologists proposed, were the first to form in the magma chamber deep below the surface, long before the others that crystallised almost instantly once the lava reached the cool of the atmosphere. Surely diamonds were just phenocrysts formed similarly, only deeper in the mantle, like this newly discovered rock that brought them so explosively to the surface?

There came a time early in the twentieth century when the unfolding story was forced to include the heavy, rounded nodules found in the diamond-bearing pipes as a key part of its explanation. They were every bit as extraordinary as the kimberlite with their ultramafic, or extremely silica-poor, composition and some of them were spectacularly rich in the same garnets as found in the kimberlite. There was no doubt that they were xenoliths – fragments of rocks quite distinct from kimberlite and incorporated in it – collected by the rising kimberlite from deep in the mantle. Xenoliths of other rocks familiar to all geologists being common in kimberlite, it was easy to visualise these heavy nodules as representing material ripped off the walls of the chimney in the kimberlite magma's violent ascent, though from a far deeper level than the crustal xenoliths.

Mantle nodules: the pale disc-shaped one on the left is eclogite, the others peridotite.

In the mid-century the first nodules were found with concentrations of diamonds far richer than in the kimberlite itself. These diamonds were as exotic to the kimberlite as the mantle nodules collected by it en route to the surface. The mystery of the source of the diamonds was well on the way to being solved.

FOR THE CONTEXT of these discoveries we need to broaden our focus in both space and time. In order to reach a better understanding of our restless planet, for many decades thousands of seismic stations around the world have been continuously recording movements in the Earth's crust. From the shock waves generated by earthquakes and tremors and picked up by these stations, seismologists can start to model the Earth's anatomy. These shock waves can be likened to light travelling through various translucent media of differing densities. They bend, or refract, as they pass from one layer to another of a different density, the extent of bending depending on the densities. And if the angle of incidence – the angle at which the beam of light, or the shock wave from an earthquake or tremor, strikes the interface between two layers – is shallow, it doesn't penetrate at all but is bounced back, or reflected. Refraction and reflection are two well-understood phenomena in the domain of physics.

Put together enough prize-winning mathematicians and seismologists with seismic data from a worldwide spread of recording stations, give them the time they need, and a picture starts to emerge of the inside of the Earth. Add to it laboratory information on the known and calculated densities of different crustal and mantle rock-types and the results of gravity surveys and other geophysical inputs, and the model gains a clarity that is remarkable and at the same time credible.

The diagram (featured here) illustrates better than any paragraphs of text what we

Plate tectonics is a theory to explain observations such as sea flow spreading, rift valleys, subduction zones and related mountain belts, major faults, and most volcanoes and earthquakes.

- *The lithosphere (outermost mantle and crust) is relatively cold, rigid and brittle and does not move easily without breaking. Faults form when rigid rock breaks and moves.*
- *Below the lithosphere is the asthenosphere (hotter upper mantle) that moves like putty as it is driven by heat-generated convection currents from the lower mantle and core. Cutting through these currents are jets of heat called hot-spots and plumes.*
- *These convection currents break the Earth's outer skin into a series of plates that then move relative to each other. Major faults form where plates slide past each other.*
- *Spreading centres, rift valleys and volcanoes form where plates split and separate; this is normally the start of new seas. An example is Iceland, an island formed by volcanoes of the North Atlantic spreading rift.*
- *Subduction zones, mountain belts and volcanoes form where plates converge and collide. The Himalayas, Alps and Andes are collisional mountain belts. Subduction recycles crustal rock back into the hotter mantle like a conveyor belt.*
- *Earthquakes occur at all sites of plate movement.*

know about the structure of the Earth, and the way plate tectonics works. It is against this background that we must view the discovery that kimberlites generated in the asthenosphere had collected diamonds, as well as quite a lot of the host material in which the diamonds had formed, from the base of the lithosphere.

TO START WITH, let us revisit that iconic field geologist of the old school, Dr Tom Clifford, who used a great store of geological mapping to make a fundamental subdivision of African geology. Clifford found that the continent consisted of ancient cratons, stable for more than two billion years, and around them were wrapped belts of younger formations that had been crumpled and reorganised, subjected to temperatures and pressures high enough to change the identity of the rocks altogether.

When Clifford plotted diamond-bearing kimberlites on the map, they were all on cratons. Even the alluvial diamond workings were on or adjacent to the cratons except where they had manifestly been transported far from their source. The scores of kimberlites punched up off the cratons, through the so-called mobile belts, were all sterile. For diamond explorers Clifford's message was simple: make sure you're on the craton.

Initially the rule was empirical. In 1966 Clifford could not spell out why kimberlites that had erupted off the cratons almost certainly would not have carried diamonds, but all the field evidence showed it to be a valid maxim. Nearly half a century later, with a clearer, more coherent conception of the architecture of our planet, we believe we understand why it works. Gem diamonds form in the rigid part of the mantle, which we call the lithosphere, directly under its surface layer, the crust. Below it the asthenosphere is mobile by comparison, though not molten. Within the lithosphere there is a lens-shaped part, beneath the continental cratonic 'keels', where the pressure is high enough – and the temperature not too high – to form diamonds and for them to grow in equilibrium with this environment. This is called the diamond stability field.

An extremely resorbed diamond, as it reached the surface.

Kimberlite magma forming well below the diamond stability field and travelling to the surface may collect en route chunks of the rock layer in the lithosphere where the diamonds formed and carry them to the surface. To these broken-off fragments of mantle rock, the kimberlite magma is aggressive, and they tend to break up quite quickly, releasing their component parts, like olivine and garnet – and diamonds – into the kimberlite. The bigger pieces may just be rounded as they tumble about in the ascending magma, and survive to be blasted into the atmosphere and fall back into the pipe. They are perfectly preserved samples of a part of the mantle 150 to 200 kilometres below the Earth's surface.

What do we mean when we say that the kimberlite magma is chemically aggressive towards the mantle fragments? At the time the kimberlite captures the chunks of the layers through which it passes, it is still in its magmatic stage and loaded with 'hydrothermal' fluids. While these juices are in equilibrium with the kimberlitic minerals that share their origin, they do not share this stability with any pre-formed 'foreign' material with which they may come into contact. Under the attack the smaller entrained mantle chunks are disaggregated, releasing their component minerals into the magma, and exposing them in so doing to the hydrothermal fluids. These start their work immediately, among other things oxidising or 'resorbing' the diamonds. If a diamond is thrown into a very hot fire – above 900°C – it will burn up to carbon dioxide; so you can imagine their vulnerability to the superheated kimberlite magma.

One of the key phrases in the diamond geologist's vocabulary is the diamond stability field. We now have a good idea of the parameters of the temperature, pressure and, by deduction, depth of that magic lens in the mantle where diamonds form. Ironically this information has not come as much from pure-carbon diamonds as from the minerals that some of them bring to the surface enclosed within them. In the magma where the diamonds form, other minerals are forming at the same time. If another mineral is growing very close to the diamond, the diamond may grow around it, ultimately entrapping it completely. For all practical purposes the two minerals share the same age and form under identical conditions. In most cases the enclosed crystal is typically quite a complex 'rock-forming' mineral made up of silica (SiO_2 or silicon dioxide) and other metallic oxides, usually including aluminium. Because of its complexity it contains a great deal of information useful to diamond geologists, such as the pressure and temperature at which it formed.

An inclusion inside a diamond.

◆◆◆◆◆

DIAMOND GEOCHEMISTS like John Gurney at the University of Cape Town found that they could remove the enclosed minerals from the diamonds and conduct all the analysis they wanted with the ultra-sophisticated equipment that has become available in recent decades. In samples of diamond-bearing kimberlite or mantle rock from which the diamonds had derived, they came across the same data-filled minerals elsewhere in the rock, not just caught up in the occasional diamond.

Gurney was more than a learned university professor. He had a passion for applying his own laboratory findings, as well as those of others, to finding new mineral deposits. As a result he formed a consultancy to assist exploration companies with rock and soil analysis that would help them home in on the bonanza which their research had told them lay below the surface of the ground they were prospecting. Gurney's all-consuming quest became the identification of a geochemical key that would unlock the secret of whether a newly discovered kimberlite carried diamonds. To do this he analysed the minerals in diamonds and the same minerals in known diamondiferous kimberlite, searching for a pattern that characterised them. The most promising seemed to be the garnets. Not only were they common as inclusions in diamonds, but the 'free' garnets in mantle nodules (those formed in the diamond stability field) were generally the most abundant indicator minerals found in the sampling of potentially diamondiferous ground.

To appreciate the meaning of Gurney's work, we need to know a little more about this key mineral. Think of garnets as a family of minerals. The commonest members are red almandine – the garnet usually seen in rings and brooches, which is relatively conspicuous in metamorphic rocks around the world – and pyrope, mostly ruby red, but also orange or purple, which is occasionally seen in jewellery and comes from rocks formed deep in the mantle. Garnet formulae are quite complicated, that for pyrope being written $Mg_3Al_2(SiO_4)_3$. The main metallic components of the mineral are thus magnesium (Mg), aluminium (Al) and silicon (Si). However, some of the magnesium will be replaced by small amounts of calcium, iron and manganese, and the aluminium by traces of iron, chromium and titanium. It is the ability to analyse the exact amounts of these various metals in the pyrope make-up that gives geochemists the clues as to the temperature and pressure that prevailed when and where they were formed.

The colour of garnets varies according to changes, often subtle, in composition.

The data for the temperature and pressure of formation of the different garnets were, to an extent, of academic interest. What was of primary importance in the immediate term was the ability to stereotype the garnets that accompanied the diamonds, either inside them or alongside them. By doing this and comparing them with garnets from kimberlite known to be devoid of diamonds, scientists could use garnet composition to indicate the probability that kimberlite would be diamondiferous. This became the Holy Grail. The question was what to use as a basis of comparison.

For a long time the task seemed hopeless. The different elements analysed for (magnesium, calcium, divalent iron, manganese, aluminium, trivalent iron, chromium and titanium) showed a scatter of values too wide to be of any use when plotted individually. No pattern was discernible. Then Gurney noticed a trend of two main elements that seemed to be consistent in the Group 10 (later shortened to G10) garnet population from the diamond-

JOHN GURNEY

More than with any other mineral, the history of diamond exploration is marked by milestones marking where its progress moves up to a new level, where the search becomes faster, easier, more focused. Such a milestone was the recognition that minerals (such as garnet) contained within diamonds had diagnostic compositions and that this knowledge could be used in a predictive way by identifying genetically related minerals in early stage exploration programmes. The discovery revolutionised the industry. And the name carved into that stone is Dr John Gurney's, as much entrepreneur as multi-published academic. Although he would be the first to say that he did it as part of a team, there is no doubt that he was the principal architect of the technology and its application, and was instrumental in collecting hundreds of samples from mines and prospects around the world and having them analysed. Now geochemistry can short-circuit years of field and kimberlite sampling. This has been recognised worldwide.

bearing kimberlites and the peridotite nodules in them. They were all low in calcium, or subcalcic, and rich in chromium, or chrome-rich. He plotted the values for these two on a graph with calcium on the horizontal axis and chromium on the vertical axis. The G10 garnets formed a close cluster. Gurney then assembled all the garnet analyses he could find and plotted them. The G10s continued to form a tight group. He drew a line separating the G10s from the others: it turned out to run across the graph in a direction roughly parallel to the hour hand when it is one o'clock.

The upshot was that garnet compositions plotting to the left of the line were cause for champagne to be popped; G9s, just to the right of the line, and chrome-poor G4s, close to the line but right at the bottom, might be worth a glass of sparkling wine (or not). If the compositions were far to the right, it was back to the field as another kimberlite was declared sterile.

The early days of the application of garnet geochemistry to the search for diamonds have been described by the journalist Kevin Krajick in his history of the Canadian diamond boom

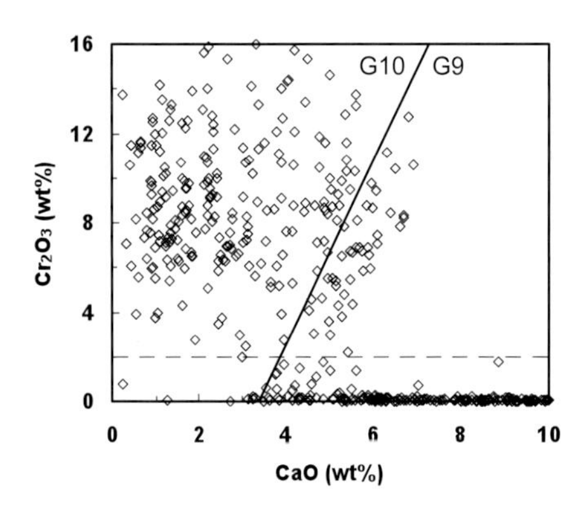

A plot of chrome and calcium contents in garnets included in diamond.

entitled *Barren Lands*. In 1978 the Superior Oil company had secured the services of a top South African-born and -educated exploration manager for its diamond exploration arm in Canada, Hugo Dummett. He in turn had signed on Tom McCandless as his assistant. Not only did Superior Oil do exploration for its own account, but it had acquired a fully fledged mining and exploration company, Falconbridge, with Chris Jennings as their southern African exploration manager. Jennings and Gurney had worked together from the start. Now Superior and Falconbridge put up $100 000 to fund Gurney's ongoing research, which seemed on the verge of a breakthrough.

According to Keven Krajick, 'Great news came from South Africa. John Gurney had found the secret indicator-mineral formula.' A few pages later, Hugo Dummett broke the news to his assistant:

'Early on, Hugo called McCandless into his office and let him in on the G10 secret. On a scrap of paper he drew the *x*/*y* axes showing the low calcium, the high chromium and the sloping line. Fearing the room could be bugged, he took care to explain it by pointing to the elements, but not speaking their names.

'"Do you think you can memorise that?" asked Hugo.

'"Sure. It's simple," said McCandless.

'"Good," said Hugo. He ripped the paper to shreds, piled them in his ashtray, and set them on fire.'

Such was the intrigue in the early years in Canada.

It might seem that the conclusion John Gurney reached was elementary. Those working for him at the time, and for the Russian Nikolai Sobolev (son of the celebrated pioneer Vladimir Sobolev), as well as geochemists for De Beers, will testify that the answer they sought was far more elusive than the narrative here might suggest. And they were all after it, independently and very much at the same time.

Over the decades exploration geology – certainly the most exciting and intriguing branch of the science – has seen a few seminal technological breakthroughs. Most of them have been in the field of geophysics: for example, the discovery that airborne instruments could detect base-metal (mainly copper, lead, zinc and nickel) ore bodies buried under the surface of the ground. The companies associated with the unfolding technology have naturally gone to great lengths to retain the proprietary advantage as long as possible. Secrecy is paramount.

But even if walls cannot have ears, somehow it never takes very long for the news of an important leap forward to leak out.

Consider the paper by John Gurney, written in 1977 for the *Industrial Diamonds Quarterly*, or *Indiaqua*, titled 'Formation of the Diamond (A Geological Whodunit)'. Referring to inclusions in diamonds, compared with related minerals in 'common peridotite', Gurney simply says, 'They have higher Mg/Fe ratios, the chromium contents are higher, the garnets have low calcium contents, and so on.' Neither in that issue nor in the previous one, where he wrote the paper 'What is Kimberlite?' is there any mention of G10s or other garnet groupings.

This is hardly surprising, since whether Gurney had made his breakthrough or was on the verge of doing so, his research had been funded by the Canadian companies already referred to and he had no licence to make its findings public. Nor, being a professional, would he have done so: it was privileged information in the strictest sense. Yet he must have been bubbling with excitement at the importance of what he was discovering. He knew diamond exploration would never be the same again.

No matter how tightly he kept his secret and how carefully Hugo Dummett burned tiny scraps of paper in his ashtray, in the world of mineral exploration good news can only be contained for so long. Within a year every diamond geologist knew the anxiety of waiting for calcium and chromium analyses of the garnets they had collected so carefully. Were their garnets G10s?

IN THE 30-ODD years since Gurney and his colleagues made their breakthrough, a remarkable amount has been learned about the formation of diamonds and their transport to the surface. This has gone hand in hand with a steady improvement in our understanding of the anatomy and workings of the Earth's mantle. We now know that the oceanic crust, which started in the eerie depths of deep-sea troughs, is being continuously driven down, initially under continental plate edges and then so deep under the cratons that it melts, the carbon in it subsequently crystallising as diamond. At the same time diamonds crystallise in the primeval mantle, some of them growing around tiny garnets and other minerals forming at the same time.

Part of the reason for writing this book, and particularly this chapter, is to illuminate how diamonds have offered a window into a world as fantastic as Alice's. Consider this title, not as formidable as the long and unfamiliar names suggest: 'Tracing Lithosphere Evolution through the Analysis of Heterogeneous G9/G10 Garnets in Peridotite Xenoliths'. To summarise some of the terms: the *lithosphere* is that more or less rigid part of the Earth lying immediately below its surface and above the more mobile asthenosphere. *G9 and G10 garnets* are Group 9 and 10 garnets, of a certain defined chemical composition, which are stable in quite restricted ranges of temperature and pressure, and whose age can be accurately determined. *Peridotite* is an ultramafic, or very basic (as opposed to acidic), silica-poor rock formed in the mantle. *Xenoliths* are fragments of rock ripped off the walls of the passageways that any magma uses to reach the surface.

A diamond contained within another diamond.

The garnets – to us mainly of interest for what they tell us about diamonds and where to look for them – have been the keys that unlocked otherwise inaccessible secrets held in the lithosphere. For scientists trying to understand the Earth, the information provided is of vital relevance. Garnets contained in diamonds have the unique ability to give geologists accurate information about conditions in the mantle because their envelope – the diamond – shields them from any attack by the kimberlite and its hydrothermal fluids. But we also find that the 'free' garnets in the xenoliths of mantle material (not held in diamond) have the same chemical composition as those included in diamond, as do those that have been released from the xenoliths into the kimberlite as the xenoliths disaggregated.

Exploration geologists need to be able to tell from garnets they've collected in samples of stream sediment – or soil or kimberlite – whether they should scale up their exploration budget or keep on as they were. This is because a lot of the exploration we're talking about may take place in remote parts of Africa or Siberia or northern Canada. Claims may have to be staked, which can involve hiring helicopters. Decisions may need to be made at a senior level in the company. It is not far-fetched to say that the calcium and chromium values of garnets from reconnaissance samples will be of key relevance to making these decisions.

Diamond-bearing mantle xenoliths are of three different main types. The commonest are G10 (low calcium, high chromium), and nodules containing G10 garnets are called harzburgite. Harzburgite (a type of peridotite) is a prominent component of the Earth's mantle, both within and beyond the diamond stability field, and is made up of olivine and pyroxene, with lesser amounts of red-to-purplish pyrope garnet. Slightly higher in calcium than G10s are G9 garnets, which are from a rock called lherzolite (another kind of peridotite), not very different geochemically from harzburgite, the main obvious difference being that they carry bright green chrome diopside, sometimes in conspicuous amounts. Lherzolite has produced a number of extremely richly mineralised samples.

Much less common than G10 and G9 garnets as diamond inclusions are G4 garnets, much poorer in chrome than the other two. The G4-bearing eclogite nodules also carry concentrations of diamonds that may be spectacular – thousands of times richer than any kimberlite – and are far commoner than diamondiferous harzburgite. Although eclogite has the same chemical composition as dull dark-grey basalt, it is very dissimilar to look at, being brightly colourful and coarse-grained, made up principally of greenish (chrome-poor) pyroxene and orange to orange-red garnet. It has a more interesting pedigree, too. Whereas harzburgite and lherzolite are part of the primeval mantle, eclogite has an altogether different origin. It formed from the basalt erupted along the mid-oceanic ridges aeons ago, as we have mentioned. So, both for what they tell geologists and for the occasional fabulously diamond-rich mavericks among them, the mantle xenoliths are the caviar for miners and earth scientists alike, hard to come by and incomparably valuable.

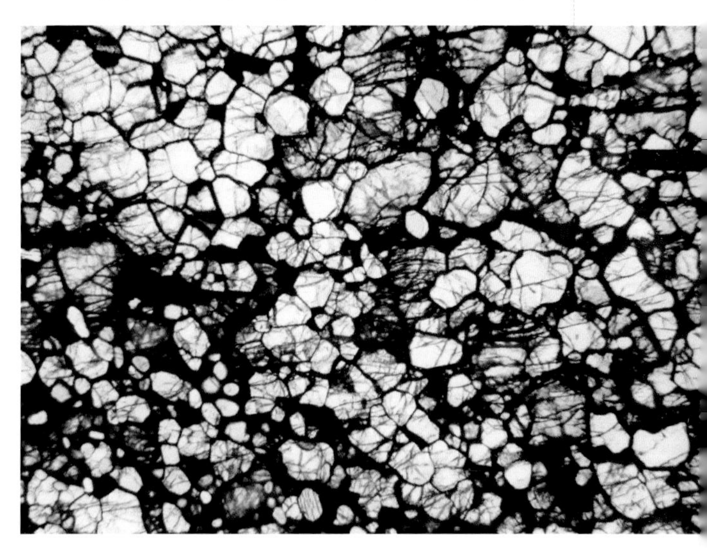

Slice of an eclogite xenolith showing omphacite pyroxene (blue-green) and garnet (pale lime green).

◆◆◆◆◆

LET US NOW look at what happened to garnets formed deep in the mantle when they became exposed to the ascending kimberlite magma. They reacted with the magma so that a reaction rim (called a kelyphytic rim by geologists) formed around the grain, protecting it from further attack. In the sorting process after mining, dull grey-green balls the size of a marble or a pea coming onto the conveyor belt with the final plant concentrate will often reveal bright red, orange or purple garnets inside them if collected and broken open. Not only did the kelyphytic rim protect the garnets from further kimberlite attack, but given half a chance it will hide them away from the waiting geologist.

At first it might seem fortunate that the diamonds did not protect, and hide, themselves in this way. In fact what happened was much worse. They might well have oxidised completely to valueless CO_2. If, however, the process had not run its full course before the kimberlite reached the relatively cold surface layers and then the atmosphere, the magma would have 'frozen' and there would have been no more attack. This has happened often enough through geological history to ensure a relatively generous supply of diamonds to areas where we can mine them.

What might you see if you could watch the kimberlite erupting, ripping bits off the wall in its ascent? The breaking down of mantle material starts as soon as it comes into contact with the kimberlite magma and continues as long as the hydrothermal fluids are active – in other words, until the magma is exposed to the cool upper crust and the atmosphere. The process continues for the entire ascent of the magma from a depth of 150 or 200 kilometres to close to the surface. By the time it vents, some of the diamonds will have been exposed to the magma during the whole ascent, while others will just have been released from the peridotite or eclogite, their true parent rock.

This is why in practically any diamondiferous kimberlite the diamonds that have not been very deeply resorbed during the ascent will show a range of shapes. Like some of the minerals geologists deal with, the crystal forms have tongue-twisting, Greek-derived names. Diamond sorters see shapes ranging from perfectly octahedral (8 crystal faces), mostly tiny, to those approaching a dodecahedron (12 crystal faces) or a tetrahexahedron (24 faces). The change in crystal form occurs because the corners and the edges, where crystal faces join, are more susceptible to removal by magmatic corrosion than the flat faces themselves, as you can see in the accompanying figure. Note how the resorption has the effect of rounding what started as a very angular shape.

Ironically the rounding is wonderful preparation for what is to come once they reach the Earth's surface. The six sharp points of an octahedral crystal give it a vulnerability during river transport that the rounder stones do not have. Not that the 'octas' become abraded during their journey among the violently jostling cobbles and boulders in flooded rivers; they are just more liable to be broken along their cleavage planes – parallel to the crystal faces –between two colliding boulders. Notwithstanding that heightened susceptibility to breakage, stones transported down the Orange River, then more than a hundred kilometres up the wild Atlantic coast, and finally into the Namib Desert by wind that can sand-blast the duco clean off vehicles in days, contain a fair share of octahedra that have survived. They take your breath away, their points are so perfect.

Diamond shapes: (from left to right) resorbed dodecahedron, octahedron, macle, cube and cubo-octahedron.

What have the diamond detectives been able to deduce about the time it takes for the kimberlite ascent we have been talking about? In a world where many of the shaping processes happen over millions of years, are we looking at millennia or years or days? Probably months, say the experts, which, in the circumstances, is quick, too quick for large diamonds to be resorbed away to nothing.

THE YOUNGEST kimberlite-type pipe – the Ellendale pipe in Australia, comprising a rock called lamproite, which is quite like but not identical to kimberlite – has been dated at 20 Ma. In geological terms that is not very long ago but in the context of historical events it is unimaginably far back in prehistory. While most of the geological events and rock formations geologists work with predate the Ellendale eruption by aeons, there has still been a lot of change to the face of the planet in 20 million years. Look at the turgid river you cross as you drive along the coast. Try to visualise how much sediment it carries out to sea in a year: tonnes, hundreds of tonnes; in a century, tens of thousands; in 20 million years, billions of tonnes, all stripped off the surface of the continents we think of as unchanging.

The point is that most of the kimberlite pipes geologists see at the surface have had tens or hundreds of metres of their uppermost part stripped off since they vented, together with the rock into which they intruded. The landscape at the time of their intrusion was very different from the one we know. Some pipes, however, like Mwadui and Orapa, still have their uppermost parts – the crater facies – preserved or did have until mining removed the material to get at the diamonds. Mining companies take no prisoners. In another kind of world we might have had such rarities preserved in a super-museum. In the real world the most we have are very complete records of what the crater facies looked like and, in the case of Orapa, a comprehensive collection of the fossils recovered as mining progressed.

Visitors to the Bernard Price Institute for Palaeontological Research at the University of the Witwatersrand can see, among many other treasures, a unique collection of fossils. The most delicate plant and animal remains in the Orapa display were painstakingly recovered from the top parts of Botswana's first kimberlite pipe as mining got under way. Together with all other aspects of that 'crater facies' kimberlitic sediment, the fossils were used by a panel of geologists to recreate the conditions when the pipe burst through to the surface 93.1 million years ago.

Here is a reconstruction of what it was like immediately after the eruption of the Orapa kimberlite volcano.[1] Around a 110.6-hectare shallow crater-lake, or 'maar' (nearly 1.2 kilometres in diameter), was set in a ring-shaped range of hills. At first the crumbly, loose kimberlite, with no clay or roots to bind it, washed into the lake with every tropical thunderstorm or was blown in by seasonal gale-force winds. Over the months, secondary

Sediment-laden rivers like this lower the land surface, slowly but surely.

clay minerals formed from the olivine and other kimberlite minerals, and seeds blew onto the hill slopes from far and wide. They took root and a plant community was born. Low bushes grew, soon to be shaded through the hot summers by leafy trees. Before long a host of insects made their home in this pristine new environment, and who knows what other fauna. Leaves fell in autumn storms and, together with beetles, spiders, weevils and aphids, were washed into the lake and buried by more mud.

The uppermost preserved sedimentary layer – after some tens of metres had been stripped off by erosion and part of the earliest mining – is 'Facies IV Fine-Grained Sediments'. It consists of 'well-bedded, fine mudstones, interbedded with mud and debris flows'. What is more relevant, and more exciting, is that it is classified as a *Lagerstätte*, or exceptional fossil locality.

There are only a handful of *Lagerstätten* around the world, nearly all in the Northern Hemisphere. They are rare because they arise from a freakish combination of circumstances. In the first place, dense concentrations of creatures must come together in a small area. They must also be encapsulated for posterity and preserved – protected down the ages from removal in the endless process of recycling. Lastly they have to be exhumed for us to find them. Only if all those requirements are met will *Lagerstätten* be left to us. Perhaps the most famous is the 530-million-year-old Burgess shale in British Columbia, described by that doyen of science writers, Stephen Jay Gould, in *Wonderful Life*, which set some basic precepts of Darwinian evolutionary theory irrevocably awry.[2]

Volcanic craters that are no longer active must be a prime candidate for *Lagerstätte* creation, and Orapa is an example *par excellence*. Along the bedding planes in the mudstones were found carabid beetles, weevils, wasps, fossil March flies, aphids and spiders, as well as 29 leaf types, flowers, seeds, fruit and ferns, all from angiosperms (flowering plants), and pollen from three gymnosperms (pine trees, cedars, cypresses, etc.). Most of the life forms are only known from this locality, so the number of *orapaensis*-specific names is not surprising. The fossil finds are a window into the mid-Cretaceous in southern Africa of extraordinary detail and preservation.

There is an excellent paper on the locality and its fossils by a group of palaeontologists mostly from the Bernard Price Institute. The excitement conveyed in the lines of the monograph is almost palpable. They describe wing venation of Diptera (a huge family of flies, midges and mosquitoes) 'preserved in exquisite detail', a 'new species of wasp … the first of its kind from the southern hemisphere Mesozoic', a spider compression 'the first to be described from Africa … with pigmentation … even preserved on the legs', and plants 'unlike anything that has been described from the Northern Hemisphere'.

The Mesozoic was an era when the dinosaurs reigned supreme, so one might have hoped for one of them at least to have shown up. But considering the fragility and delicateness of what was preserved, the absence of larger fauna is hardly surprising. There is, however, circumstantial evidence that they may have been around. The paper notes, 'The presence of a possible dung beetle in the absence of other fossil evidence for its large herbivore benefactors [to provide appropriate dung] invites the suggestion that they were associated with large dinosaur herbivores in the mid-Cretaceous.' There are other fascinating implications for the pre-Continental drift association of weevils and proteas in Africa and Australia, and valuable insights into climate at the time. Southern Africa's own *Lagerstätte* has undoubtedly provided a uniquely important insight into our past.

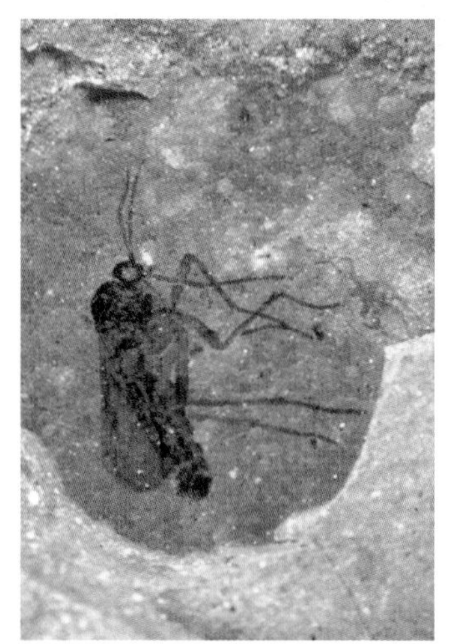

Most of the miraculously preserved tiny insects from Orapa are not known from other fossil localities.

Not far below the fossils, scattered in the coarser-grained sediments, would have lain diamonds – carbon of a very different kind from the flora and fauna that settled out of the lake water onto the mud layers. They are from opposite extremes: living organisms requiring extraordinarily sensitive circumstances for their preservation, and the diamonds capable of survival in all their perfection in defiance of the most violent sequence of environments imaginable. But, we ask ourselves, was some of the carbon used to make the diamonds not organic in origin? Orapa has a high component of eclogite xenoliths and that eclogite is recycled oceanic crust. Was there not some primeval organic ooze on the ocean floor at the time, which was subducted to form part of the melt? Probably not at Orapa, geologists say, but the possibility of its happening elsewhere has not yet been discounted, and some kimberlite geologists argue that it was very probable. If so, can there be a more extreme example anywhere of the transformation of material from one form to another with no significant addition or subtraction?

On the assumption that it did happen, a moment's thought will tell you that it must have taken many millions of years to get from a starting point of micro-organisms living in the open ocean to diamonds crystallising at 40 000 bars pressure and over 1000°C, 150 kilometres or more below the surface of the Earth. Compare that infinitesimally slow process with another diamond-forming event, when an asteroid struck Earth in northern Siberia at Popigai around 36 Ma ago. The impact at Popigai may not have extinguished a major family like the dinosaurs, nor is it as big as that at Vredefort in South Africa, but it is by far the best-preserved and best-displayed mega-impact crater anywhere in the world. Its status, however, is destined to remain scientific. At 71° 38'N, not only is Popigai far north of the Arctic Circle (66° 33'N) and extremely difficult to reach, but it consists of by far the biggest single deposit of diamonds in the world, and the Russian government is unlikely to be offering midsummer excursions there any time soon. Nor, ironically, is it likely to be looking to Popigai as a future source of industrial diamonds, which is what they are. (Unlike kimberlite and normal alluvial deposits, it contains not a single gem.) It is cheaper to manufacture industrial diamonds in the cities, and to recover them at practically zero cost from gem mines, than to open a mine far further north than the already near-polar existing Russian mines. But as a curiosity Popigai has more than earned its place in these pages.

The peculiar 70-kilometre circular depression with its extraordinary assemblage of rock-types had first been noted in the mapping of the Siberian platform in the late 1940s. It was space and lunar exploration 25 years later that asked the question whether there had been the same intensity of asteroid assault billions of years ago on Earth's surface as there had clearly been on the moon. If so, might Popigai be such a crater?

Every worthwhile project needs a champion; and the champion of the Popigai crater and its geology turned out to be the Russian geologist Victor Masaitis, whose papers have provided the material for our story. It is thought that the 'impactor' was a chondritic (stony) asteroid (very large meteorite) only slightly denser, at 3.5 g/cm^3, than the mantle rocks already mentioned. It would have been about 8 kilometres in diameter and is estimated to have struck the Earth at a speed of 20 km/s or 7200 km/h. On impact the 'target' rocks reacted in a variety of ways, depending on whether they were at or below the surface, and how far they were from the centre of the point of impact.

Around 99 per cent of the impactor itself vaporised on contact, the remaining 1 per cent being incorporated into the melt created from the target rocks that were not themselves vaporised. The evidence in the field leaves no doubt that thousands of trillions of tonnes of rock were melted, about half staying within the crater, the other half being ejected. What was not melted was shattered, the resulting blocks being thrown in every direction, downwards,

A section through the crater filling at Popigai. Any bomb blast is trivial by comparison.

outwards and upwards, the last-mentioned retaining enough of the impact energy to leave at supersonic speed, some of them landing 70 kilometres beyond the crater. A plume not unlike those over Hiroshima and Nagasaki in 1945 would have been seen from tens of kilometres away and would have comprised solid, molten and vaporised material, in some respects resembling the plume of an explosive kimberlite eruption, except for its much greater scale.

The discovery of the diamonds at Popigai was accidental. During the sawing of rock samples for microscopic investigation, a case occurred where resistance was encountered and proved too hard to be sawn through. It could only be a diamond. Closer investigation showed that there were other such grains, mostly yellow or grey.

It was known that the gneiss beyond the limits of the impact effect was graphite-bearing. (Like diamond, graphite is pure carbon.) The graphite in the gneiss must have converted instantly to diamond with no change of habitus or shape. In turn the graphite grains were assumed to have metamorphosed from biogenic carbon in the shale that was the precursor to the gneiss before 2.4 billion years ago. Mostly the gneiss contained less than 1 per cent of graphite, but in places it reached over 5 per cent.

In the centre of the impact area the temperature was too hot for diamond to form or, if it did, it reverted almost instantly to graphite. In the same way as the mantle contains particular diamond stability fields, so there was a critical part of the target gneiss where temperature and pressure were right for the conversion of graphite to diamond. This was between 12 and 13.6 kilometres from the centre of the crater in a hemispheric crescent. Imagine the amazement of Victor Masaitis and his colleagues when they mapped out the diamond-bearing zone and considered what had happened.

Laboratory follow-up at an early stage showed that the diamonds had indeed retained the crystal form of the graphite from which they had formed. Graphite belongs to the hexagonal crystal system, whereas diamond is isometric or cubic. According to the classic *Dana's Textbook of Mineralogy*: 'In graphite the carbon atoms are arranged in parallel layers, those of each layer forming interlocking hexagons … the distance between successive layers [is] unusually large [which is why] graphite cannot be readily powdered but separates rather into minute flakes … which makes it such a good lubricant.' Obviously this quality does not apply to the graphite-derived diamonds at Popigai, but we can start to see why they could not be used for gems.

Artist's impression of a meteorite impact as seen from space.

IN SOUTHERN SPAIN and across the Straits of Gibraltar, in the Atlas Mountains of Morocco, there are places where geologists travel to see one of the wonders of the world of diamonds.

The western Mediterranean coast is shaped like an archery bow drawn to release its arrow, the Straits of Gibraltar being where the arrow rests. Imagine these two localities at either end of the bow. Geologically they are as closely connected as the two points of a bow. In the African locality, graphite crystals are found consisting of up to 15 per cent of lenses of a rock closely resembling the lherzolite xenoliths in kimberlites, some of which are profusely diamondiferous. The formation enclosing the lherzolite, also mantle-derived, is non-graphitic, an occurrence reminiscent of the nodules in kimberlite in which, with rare exceptions, most are not diamond-bearing. Everything about the graphite crystals suggests that they formed as diamonds. They have octahedral and other shapes characteristic of diamonds as opposed to rhombohedral, which is the normal crystal system for graphite. They have growth features seen on diamonds and etched trigons, almost a defining attribute of diamonds. What happened, then, to devalue them into a mineral used for purposes as mundane as ensuring smoother-running engines and filling millions of pencils?

The answer is complex, and what is given here is just a summary of what geological supersleuths using nanotechnology have deduced from a long and convoluted chain of clues. Diamondiferous mantle material started its slow ascent from the depths (150 kilometres) during the plate collision that formed the Alps, beginning 85 Ma ago. The main uplift happened as recently as 20 Ma ago, when slabs brought up into the crust became detached. Even then the mantle rocks were not exposed at the surface. This only happened when the formations overlaying the mantle slabs were stripped off at a great pace by erosion from a pile of continental lithosphere that had become 'overthickened'. Nature will always try to restore the balance.

◆◆◆◆◆

TO CLOSE THE chapter, we turn our thoughts back to the asteroid hurtling through space over northern Siberia. We stay in Russia but move to the south, to a village called Novo Urei, 530 kilometres east-southeast of Moscow. On 4 September 1886 the residents of this rural spot were startled out of their tranquil existence in a way that none would forget. Here is a first-hand account of what was experienced, written by a teacher in the village school: 'In the morning several peasants ploughed their field 3 kilometres from the village. The day was gloomy, the whole northeastern sky was covered by clouds. Suddenly a light appeared all around. In several seconds a strong report was heard, like a cannon or explosion. Then came a second, louder noise. With a loud noise a fireball fell to Earth a few metres from the peasants. Frightened, they did not know what to do. They fell to the ground and could not move for a long time. They thought it was a strong thunderstorm, and that thunderbolts were falling from the sky. Finally, one of them, more brave, came to the place where the thunderbolt had fallen, and to his surprise found only a shallow hole. In the middle of the hole a black stone lay half-buried in the soil.'[3]

The town of Ronda, in southern Spain, gives its name to the nearby ultramafic massif which contains graphitised diamonds.

The meteorite is singular not only because it was the first where diamonds were noted, and is much more bizarre: it is the only recorded case where those who observed its landing collected some of the material and ate it.[4] The reasons for this extraordinary behaviour were not recorded and one can only speculate. Did they suppose it was manna from heaven?

The fragment that made its way to the laboratory in St Petersburg was found to have a composition not unlike mantle material (mainly olivine and pyroxene), without the garnets but with microdiamonds. The Mordovian village has become immortalised in that this class of meteorites is now known as ureilites.

In fact the carbon in ureilites, like that in the Popigai target gneiss, started as graphite and was instantly converted to diamond. Ureilite diamonds are, however, distinctive in the sense that they are a sample of material from outer space – even if in space it exists as graphite.

◆◆◆◆◆

THERE ARE in fact respected diamond geologists who argue that diamonds from outer space have found their way to Earth. In the Central African Republic and deep in Brazil, carbonado diamonds, black and charcoal-like except for their greater weight, have been found over the years. In fact the biggest diamond ever found, marginally bigger than the celebrated Cullinan from the Premier (now Cullinan) Mine near Pretoria, was found in Brazil, ironically in the same year as the South African stone, 1905. It weighed 3167 carats (compared with the Cullinan's 3106). But like the dogs on the whisky label, the two stones are black and white, as extreme in their shade as it was possible to be. And as if to emphasise the difference, the Brazilian stone was as dull as the Cullinan was lustrous, and their values were about as far as east is from west.

The fact that true carbonados, some of them very large, have been found mainly in two countries on opposite sides of the wide Atlantic Ocean was part of the early argument for their extraterrestrial origin. In 2006 a scientific paper concluded: 'The 2.6–3.8-billion-year-old fragmented body [that gave rise to the carbonado diamonds] was of asteroidal proportions.'[5] (Asteroids are defined as 'celestial bodies with diameters between one and several hundred miles'; in other words, they are big.) As this estimate of time, even the younger age, puts the asteroid into the *terra firma* of central Gondwana long before the first cracks presaged the break-up of the supercontinent, it may provide yet another clue in the unravelling of the history of the Earth. (It should be added, in parenthesis, that there are leading diamond geologists who dispute the cosmic genesis of carbonado.)

For as long as nanotechnology has existed to open up a world we will never see, minute diamonds have been found in meteorites. They are far smaller than the speck of dust that floats weightlessly in a beam of late afternoon sun, but not too small for some of them to contain in their crystal lattice traces of the rare 'noble' gases: neon (Ne), argon (Ar), krypton (Kr) and xenon (Xe). Like most diamonds they also contain nitrogen but in comparatively copious amounts. Both the noble gases and the nitrogen and carbon isotopic make-up have left scientists in no doubt that these meteorite-hosted nanodiamonds are pre-solar in origin: they already existed in the universe before our solar system came into being.

At the other extreme, outer space gives us BPM 37093 (also known as the 'Diamond Star' or 'Lucy'), a crystallising 'White Dwarf' two-thirds the size of Earth and 50 light years away. It lies in the constellation Centaurus and can be seen best in the Southern Hemisphere from March to June. The reason for its nicknames is that it is thought to comprise crystalline carbon in the same isometric crystal system as diamond. Because of its enormously greater density and a number of other differences, however, the star is not a diamond at all, or at least a very different kind of diamond from the ones we know.

A carbonado diamond.

All the same, the minuscule pre-solar diamonds out there are real enough. Even if they can't be seen with the naked eye, there are instruments that can pick them up in a piece of rock on a microscope stage, so they are a palpable reality. Not being able to see them is only part of the problem. How capable are we of imagining something that reaches us from a time long before Earth was formed?

Stand on the Witwatersrand on a clear evening or in Yellowknife or Novosibirsk and look up. As you do, feel your feet firm on ancient rock. Above you, below you, this moment, aeons ago, billions of years hence: diamonds. How perfectly this simple crystal of carbon symbolises creation in all its magnificent splendour.

CHAPTER THIRTEEN

ACROSS THE SURFACE

The diamonds at De Beers' Daberas alluvial mine have travelled from inland South Africa and very nearly made it to the Atlantic.

For millennia all known diamonds have come from alluvial deposits. From earliest times miners and prospectors could see that they were deposited by rivers as a tiny part of a chaotic load made up of sand, pebbles, cobbles and even boulders. Until the early 1900s, when the first marine and wind-blown deposits were discovered along the Atlantic coast of Namibia and South Africa, all diamonds came from river-borne sediments, ancient and modern. The wanderers of the diamond world were found where water, wind and ice had carried them to their final resting place.

For many hundreds of years, where they had started their journey was a mystery. It was only late in the history of diamond mining that the seemingly mythical mother lode was found. And although the world's major diamond mines are located in kimberlite pipes right around the globe, river gravels from as wide a geographic spread have made their contribution. Many of the most celebrated diamonds of all time – the Great Mogul, the Kohinoor, the Hope, the Dresden Green and others too many to name – were found in river gravels in central India and Brazil, while the most fabulous concentrations of gems ever discovered were found in potholes along ancient channels of the Orange River in South Africa. Along the same river not very far upstream, the Oena Mine proved a trove that for the first five years of its life produced stones of the highest average value of any mine in history. And as recently as 2002 three 'D'-colour, flawless stones of 214, 125 and 98 carats, worth around $15 million today, were recovered from a stream in Lesotho within a few days.

In its way the geology of the alluvial deposits – offshore, on beaches and desert plains, along rivers and in glacial ice-melt deposits – is every bit as fascinating as the story of their genesis.

For the billions of years of its existence the Earth's surface has been constantly in motion. Cratonic nuclei have formed in the continents and coalesced; oceans have grown as lava oozed up great central cracks that reach down into the mantle, pushing the sides apart, on and on; and other oceans have shrunk as the edges were pushed under the advancing continental plates.

Below the thickening cratons, the mantle has been forced down to form deep keels. Embedded in them, the primordial carbon in the molten mantle has crystallised as diamond, along with garnet, olivine and other mantle minerals. From the top of the cratons from time to time, deep clefts penetrate the crust, through the upper mantle and into the mobile asthenosphere, the lowest part of the mantle, which, with its porridge-like consistency, does not fracture like the more rigid lithosphere above it. At these depths the unimaginable pressure forces kimberlite magma up into the deep-reaching clefts. In its explosive ascent to the surface, the magma traverses diamond-bearing layers and tears off fragments of the rock it passes through. This is how diamonds were delivered to the Earth's surface.

As we have seen, the fiery ascent from the depths is only the beginning of a diamond's journey. What happens to it once it reaches the surface? Most of those blasted out of the neck of the pipe fall back into the crater and lie there for millions of years, as the process of erosion strips off the crater rim and works its way down into the body of the pipe. Within seconds of the eruption, which left in its wake an empty black chasm, most of the coarser material blasted out fell back into the neck, filling it in a matter of minutes until it became a relatively shallow depression. On the other hand, material (including diamonds) whose trajectory is not quite vertical made landfall at and away from the edge of the pipe's neck to form a low ring-shaped ridge around it.

Let us say the climate at the time is wet. Most of the kimberlite minerals quickly decompose to form a variety of clays. Only the most chemically stable – diamonds, garnets, ilmenites, chrome diopsides – retain their identity and bounce, roll and slide down the slope in sheets of water as summer storms lash the terrain. Some reach small permanent streams in the first storm, some in the tenth. Imperceptibly the ring of kimberlite hills around the crater is worn down, exposing the crater lake – itself now filled with sediment and dried out by the sun – to the regional erosional pattern. As the crater sediments and those beneath are stripped away, the diamonds are released and washed into the surrounding streams. By now, we are talking not of the tenth storm or the hundredth, but the millionth and ten millionth.

Geological detectives have reconstructed an entire kimberlite pipe system, from top to bottom – or from crater to pipe and finally dyke (what South African miners still call a fissure). They are able to do so because the same facies, with their consistent geological characteristics, keep cropping up, from one mine to another – in some mines near the top, in others part way down, and in yet others in the very depths. Because it is not economically feasible to work diamond mines over a vertical depth of more than a few hundred metres, no single mine has all the facies represented, from crater to root. Nevertheless, most of them have at least two, so it is possible to draw an idealised kimberlite pipe. Using the surface dimensions of the pipe and analysing which facies occurs at ground level, one can say approximately how far above the present ground surface the crater would have been.

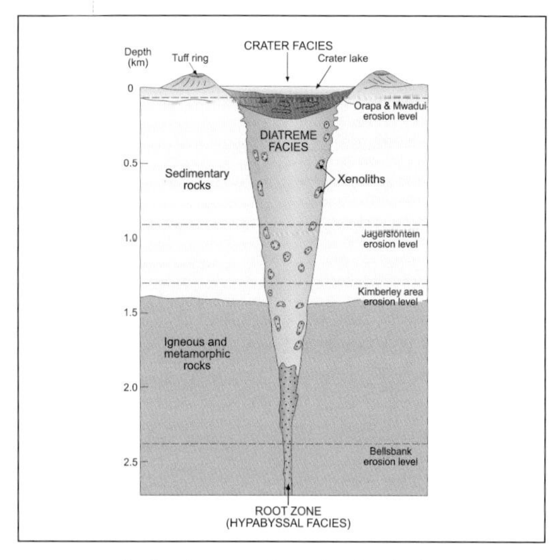

An idealised section through a kimberlite pipe, showing the different facies.

The Koffiefontein Mine sidewall illustrates the different strata penetrated by the kimberlite pipe that was mined there.

There is another way of estimating a minimum depth in the original kimberlite pipe, which is represented at the present ground level. Geologists can guess how much of the pipe has been stripped off if they can identify blocks of wall rock that fell into the pipe from above when it was intruded, and if they know how much higher the formation represented by those blocks was than the formation currently at the surface. If, for example, the wall rocks around a pipe are – in a typical South African case – fossiliferous Beaufort Group sandstone and we were to find blocks of undoubted Cave Sandstone in the pipe, we'd know we were at least, say, 1500 metres below the upper part of the pipe. This is because regional mapping has shown that the base of the Cave Sandstone (not a very thick formation), where it is found in the hills on the horizon, is that distance above the particular fossiliferous zone in which our pipe occurs.

Knowing the approximate depth of exposure of all the known pipes in southern Africa, geologists can at least hazard a guess at what caratage of diamonds has entered the total 'river budget' over time. In 1972 an article was published with the title 'The missing 3 000 000 000 carats of diamonds: New light on a great geological mystery'. Either this was an ambitious teacup reading or the author had his tongue firmly in his cheek. Since then, continuing attempts have been made to quantify how much of the gross national kimberlite diamond quotient has been 'mined' by nature, particularly in a richly kimberlite-endowed country like South Africa, which has been deeply eroded, and the guesses are tantalising.

◆◆◆◆◆

THIS CHAPTER sets out to explore what has happened to the huge quantity of diamonds distributed over the surface of the Earth over millions of years. We could make that billions, but we shall avoid hyperbole and restrict the discussion to relatively recent processes whose clues are still more or less intact. Then it becomes reasonable to extrapolate the exercise to earliest times, as long as we understand that the further back we go, the more speculative it becomes. (We shouldn't forget that the oldest diamonds found in any quantity were in the reefs mined for their gold on the Witwatersrand, rocks dated at 2890–2820 Ma, nearly 3 billion years old. The diamonds were older.)

Southern Africa has been the perfect laboratory to study alluvial processes. Although by sub-Saharan standards it is the dry end of Africa, in its heartland there are high-rainfall catchments that feed the big river – the Orange, now renamed the Gariep – which harvests its diamonds from kimberlite pipes and dykes in the dry country and deposits them along its bed lower down. In summer the river can be a torrent, deepening its gorges centimetre by centimetre, year after year.

The gravel on De Kalk, the mauve-pink pebbles with white inclusions (seen in the inset) of basalt from the Drakensberg, proving that this is a river gravel and not derived from the underlying tillite, which is older than the basalt.

In their precipitous walls cross-sections can be seen of what the river has cut through. First, at the top, you see its own diamond-bearing gravels deposited way back in the geological past, older as one goes down in the section to the top of the bedrock. This ancient substrate may have been intersected to a depth of many metres before the present river course is reached. Here, for most of the time and along most of its length the great river flows languidly.

This is the view standing at the top of a gorge on what is called the Middle Orange, somewhere between the alluvial diamond-mining towns of Douglas and Prieska. How is it possible, you may ask, for the gravels to be older as you cast your eye downwards from the top of the opposite cliff face, and yet below you, with the water lapping at its edge, is a bank of loose gravel that might have been deposited by a flood only months ago? That is certainly the youngest.

Like all rivers, the Orange both aggrades and degrades. A river is in an aggradational phase of its evolution when the principal activity is the deposition of sediment, including diamonds. The reverse process is degradation, wherein the river is cutting down. Sections of rivers flowing lazily over flat plains with no appreciable slope are aggradational; down steep gradients, they are degradational.

In the slope opposite us the gravel itself was formed during aggradation, but the reason that we see a perfect slice through it, as well as the bedrock below, is that now the river is degrading. Below us, the bar of loose gravel – very probably containing a few diamonds – deposited by the last flood reflects a temporary aggradation and is only there because the river bed has practically no gradient on it.

Rivers are not the fixed features of the land surface that you might have assumed. In the span of a lifetime the change is negligible; in geological time, though, counted in millions of years, rivers move. This accounts for beds of gravel many metres above the level of today's river, some in the banks we see flanking it, others out of sight of the river that deposited them millions of years ago and tens of metres higher. But the most eloquent testimony of all to the dynamic world we live in is rivers that have moved kilometres across the landscape, and others that have come and gone leaving barely a trace.

This gravel is far from the river that deposited it.

Let us look at the way the crustal architecture of South Africa was evolving around the time the Letšeng-la-Terai kimberlite pipe was emplaced. To the east, south and far west, new oceans with virgin coastlines had formed. What had started as a vast, gently rolling lowland plain was now rising. Earthquakes that shook the surface as faults to the east and south lifted the land. Rivers ate their way back from the coast, deeper and deeper into the interior. Where the coastal rocks were hard, the rivers cut narrow gorges, which widened imperceptibly; where they were soft, broad valleys formed. Gradually the steep slopes receded as low coastal plains established themselves. Millions of tonnes of new sediment were stripped off mountains and foothills and poured into oceans on all sides.

Up went the land – unevenly, noticeably more in the east than in the west, so there was a general tilt to the west. Back and down cut the rivers, slowing where they found a hard layer. Above thousands of knick-points temporary subsidiary base levels were formed, right across the hinterland. The ultimate base level was along the coast, always reaching back.

Where the gradient was low meanders started, first as gentle curves reflecting unevenness in the ground being traversed; then, as the process matured, exaggerating until they became oxbows. As long as they were not cutting actively down, rivers migrated sideways while the equilibrium was maintained. Sometimes the process of incision reverses, not so much because the land sinks as that the sea level rises, mainly in response to polar ice-cap melting. Along the coast, mouths and littoral sections of rivers are buried by sediment, where they are below sea level. Above the shoreline, the valleys become 'drowned' as the new, higher base level is established and the valleys fill up with sediment. They have switched from being degradational to aggradational.

Considering that all this was known to sedimentologists long ago, let us look at what secrets of landscape evolution have been unlocked by diamonds. The story starts not along the rivers but on or near the beach, close enough for the sound of the Atlantic surf to have been a never-ending distant drumming, the diamond diggings smothered more mornings than not by dense coastal fog.

After their first discovery, the extent of the coastal diamonds unfolded slowly. Starting with the fortuitous discovery of diamonds near Lüderitz in German South West Africa in 1908, there was a steady advance of diamond mining southwards until war broke out in 1914. Over a decade later, in November 1925, Jack Carstens found a diamond on the commonage of Port Nolloth. This was south of the Orange River, in Namaqualand, South Africa. A shallow pit showed diamondiferous cemented shingle, not very different from that along the Vaal and

Dr Hans Merensky (left) with associates in Namaqualand.

Orange rivers around Kimberley. That stone and its source gravel would start one of the great diamond rushes of history.

The place appeared to be in the course of an abandoned river, though in the days that followed, the great majority of diamonds were found in marine deposits, far from any fossil river. Ironically, Carstens's discovery took place in the middle of the stretch of shoreline between the mouths of the Orange and Buffels rivers, a strip that would prove the most fertile diamond alluvial mining terrain in South Africa. A frenzied search followed in two directions, north and south.

The early miners tracked diamonds with the zeal that is the hallmark of the true prospector. When they got to the rivers at the northern and southern ends of the diamond-bearing coastal gravel, they followed them upstream. For a while the diamonds and their shingle continued, though they became patchier and more difficult to find.

Along the Orange River, it was tempting to link the gravel the miners found on terraces flanking the river with closely analogous material far upriver, where the original discoveries had been made 60 years earlier. Already, while a few said that the Lüderitz diamonds must have come from kimberlite pipes at sea, most geologists had connected them with those found in the heart of South Africa.

The Buffels River and the rich fields closely clustered around its mouth at Kleinzee were a puzzle, though. Where had they come from? The riddle seemed solved, for a time, when kimberlite pipes were found in the headwaters of tributaries of the Buffels. In eager anticipation pipe after pipe was tested. One after the other was found to be sterile. Disbelief vied with disappointment in sorting rooms in the makeshift prospecting camps. The explanation for the Buffels River diamonds had gone up in smoke.

For decades diamonds kept turning up in places where conventional wisdom said there would be none. There were gems among them that proved their worth to their owners, over and over. Much greater was their scientific value. These are the little clues, the seemingly unconnected shreds of evidence, which, added together and placed in sequence, solve mysteries.

We have said earlier that geographical conditions in this South African scenario make it an almost perfect laboratory to study the processes involved in landscape evolution. Add to that its rich diamond endowment and several generations of world-class scientists, and it is hard to dispute that South Africa is the logical choice.

The unfolding diamond saga takes us to a part of the country where few foreigners have been, very near the middle of the Northern Cape. The coastal part is called Namaqualand, the eastern inland section Bushmanland. Before the hardiest white settlers ventured here, these wide plains were home to the Nama, a branch of the pastoral Khoi people, and the hunter-gatherer Bushmen. These were the primordial Africans, truly indigenous, well adapted to conditions we mostly find prohibitively harsh. Few descendants remain.

Without the windmills life in arid Namaqualand would be impossible. It was not always so.

It's a land of wide horizons and the brightest stars at night, mostly featureless and very, very dry. Shifting sand is rare, so it just escapes being a true desert. The most conspicuous element of the landscape is the huge 'pans' – in South and Central America they call them 'salars', in the United States 'salt flats', like the famous one at Bonneville. These dry lake beds are extremely flat and may reach tens, more rarely even hundreds, of square kilometres in extent. The farms are bigger, some much bigger, and you would be lucky to see the sheep that graze on them. Widely scattered wind- or solar-powered pumps deliver a thin trickle of brackish water. It's not a part of South Africa that's touted in tourist brochures, and many of its inhabitants have never seen a tree-lined flowing river. If this pan country challenges those who must make a living from it, how much more it confounded the early diamond miners.

The first diamond to be found here was by prospectors searching for calcite on the farm Paarde Kolk near the town of Brandvlei in 1937. Since then, De Beers has explored the area thoroughly and smaller operators have mined intermittently with varying success. Maintaining a viable mining operation seems a dream too far. Equally tantalising to begin with was the big question: where had the round pebbles – river pebbles if there had been a river anywhere near – come from, and the diamonds? For a long time the questions remained unanswered.

In the meantime geologists crisscrossed the area, black and white aerial photographs always to hand. Soon the topography was familiar to them: the low rises, the barely perceptible dips, and each pan with its own very distinct character. Pits were dug and boreholes drilled. How widespread were the diamond-bearing gravels? Valleys that had at first seemed subtle became conspicuous features in their thinking. Biggest of these was one they came to know as the Koa Valley, which they followed northwards, across the main road linking Johannesburg and the Namaqualand metropolis of Springbok, and on into the Orange River valley and the river itself. The Koa and its headwater stretches form a 300-kilometre-long valley, telling of a river that in its time was sizeable by any standards.

The easy part of the question had been answered – where the river went to. Much more important, and more difficult to fathom, was where it had come from, to collect diamonds in its upper reaches. Years before these questions were asked, in 1910 the field geologist extraordinaire Alex du Toit had noted jasper pebbles in gravels north of the village of Britstown in the eastern Karoo. (Jasper is a very hard rock, almost pure silica, which will survive where most others slowly decompose.) Not only were they far from any outcrop of jasper; they were a great distance from any major river. In order to resolve the problem, and to accommodate other puzzling evidence he had noted, Du Toit proposed that the Orange River had deposited them there in the deep geological past. From anyone but the revered Du Toit it might have been dismissed as outrageous: as it was, geologists nodded sagely and said, 'So be it.'

MIKE DE WIT

Most diamond geologists specialise in kimberlite or alluvial deposits, in pure research or in its application. Mike de Wit has not only worked in all fields, but has mastered them, during decades with De Beers and later as an executive in the high-risk, high-reward world of junior companies. After years as general manager for Africa exploration for De Beers, Mike was general manager of its Congo operations, based in Kinshasa, before leaving to head a Toronto-listed junior diamond exploration company. The research for his doctoral thesis, into the diamond-carrying drainage systems of western South Africa, is a seminal work in its field, explaining the occurrence of diamond fields at the mouths of rivers which today do not drain diamond source areas. Applying creative imagination to fieldwork which he carried out himself and supervised, Mike has given us a glimpse into African landscapes deep in the geological past, drastically different from the ones we know. It was a landmark study.

There was another clue: well after the mystery of far-flung diamonds had taken shape, oil-drilling off the west coast of South Africa disclosed a thick deposit of sediment off the mouth of the Olifants River. The volume of detritus was quite disproportionate to the size of the river, which as any map will show has a limited catchment area. Where had the sediment come from?

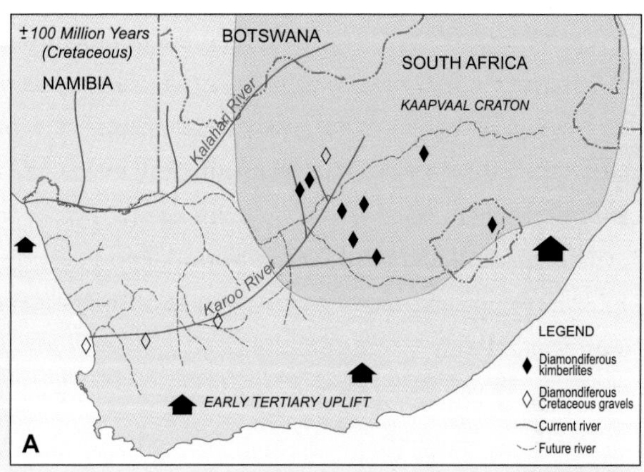

Soon after the Gondwana break-up gave Africa its own coastlines, and with accentuated uplift in the east of the young subcontinent, the Karoo and Kalahari rivers have started cutting their way through the vast plateau of Karoo basalts and the soft sediments beneath them. With most of the kimberlites younger than the Karoo and its basalts, diamonds are shed into the proto-rivers.

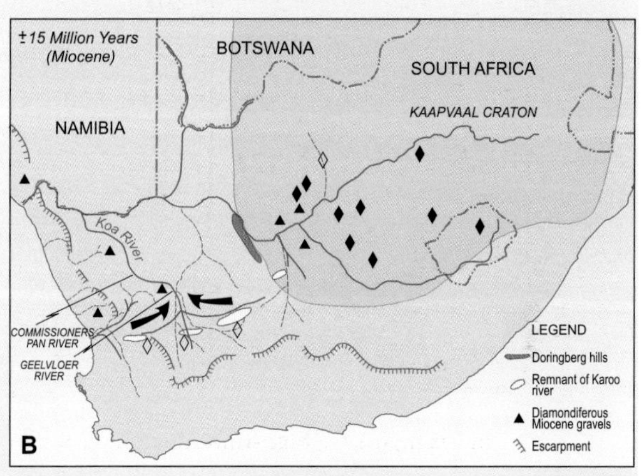

With the Karoo sediments largely cut through, the pre-Karoo topography is exhumed and starts to exert its influence on ancient river courses, substantially modifying them, and redistributing the diamonds. Here and there along the palaeoriver courses, gravels (in places fossiliferous) and a few diamonds remain as testimony to the earlier regime.

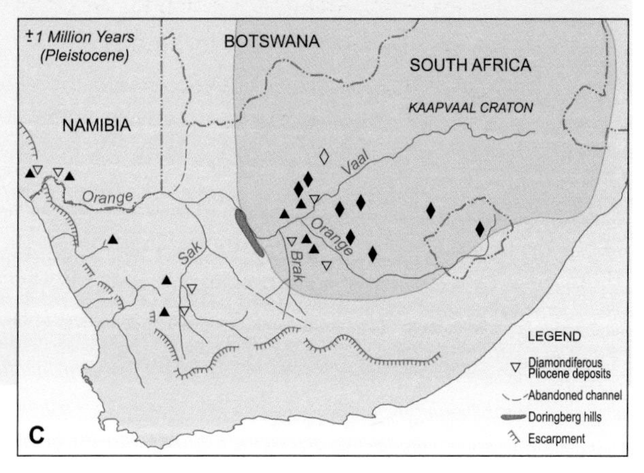

Not much has changed in the last million years, compared to earlier times. Note that the Koa, previously a major river but now choked with its own sediment, has disappeared, together with others, signalling a change to a drier climate across southern Africa. On the plains their valleys are barely perceptible.

About the time the offshore drilling was starting, a biologist made a finding that became a linked clue. Working on the distribution of fish in South African river systems, he found an unexpectedly close association of species in the upper Orange and Olifants rivers. This must have seemed bizarre, for the two rivers are separated by a wide span of some of the driest country in South Africa. Now we realise it's not so bizarre, thanks to the work of the De Beers geologist Mike de Wit.[1] For his PhD, De Wit assembled all the clues, including a mountain of proprietary De Beers geological data and years of his own work, to come up with a synthesis that worked. What emerged was a fascinating picture of 140 million years of varying climate and fluctuating river activity, to produce an ever-changing pattern of drainage networks and diamond supply.

During degradational phases young rivers cutting back to extend their headwaters may capture major catchment basins of adjacent systems, leaving the downstream part of the older river an amputated remnant.

For a lot of their early history, the inland sections of the rivers flowed over Karoo sediments, horizontal and mostly soft. Long stretches of local base-levels would have been established where there was practically no gradient. The slightest uplift along structural axes, in response to activity in the mantle deep below, would have tilted surfaces, and rivers would have migrated in response to new gradients. Once the sub-Karoo basement was reached in

Part of the CDM operation at Oranjemund: note the riffled and potholed bedrock, ideal for concentrating diamonds, swept as clean as your living-room floor.

the slow incision, the very ancient (pre-Karoo) topography – ridges where the rocks were harder, valleys were softer – would have exerted itself, changing flow directions, in some cases drastically. Aggradation, such as happened along the Koa, choked rivers up, consigning them to geological history, while new rivers cut back to pirate headwater sections of their neighbouring valleys. And on it went, and on it goes.

The two main west-flowing rivers that formed after the Atlantic Ocean had started to open up, which De Wit called the Kalahari and Karoo, mutated and their successors mutated, and theirs too, until we reach the landscape we travel across today.

Like the *lagerstätte* at Orapa, with its sumptuous collection of flora and fauna, there is a kimberlite crater in Bushmanland where a large-diameter diamond exploration borehole collected crater-lake sediments with fossils of frogs, fish, a variety of shells and a lot of wood. With no diamonds in the pipe there was never a mine there and no *lagerstätte*, though it was valuable information nonetheless. The crater sediments, as well as an abundance of fossils from some of the prospecting pits, gave Mike de Wit vital clues as to past climate. For example, a mastodon molar was found in a pit, as were bones of an early giraffe, bovids (the family that includes antelopes and buffalo) and a rhinoceros relative, as well as fragments of crocodile teeth.

If you have been in the area you will appreciate what a mind-stretch is required to imagine these creatures there. Anything remotely giraffe-like where even knee-high shrubs are today few and far between seems like a tall order. Suffice it to say that the assemblage indicates a well-watered woodland setting, steamy and stormy. More to the point, the type of environment reliably dates the fossil-bearing beds as early to mid-Miocene, between 20 and 10 million years ago. At other times, both before and soon after, it was as it is now. A hundred million years ago it was humid, with strongly flowing rivers carrying material from decapitated kimberlite pipes across the subcontinent and into the sea.

Artist's impression of a diamond and its travelling companions on today's beach at Oranjemund.

What happens then? Again we have a unique opportunity to study practically the only marine diamond placer in the world – certainly the biggest by orders of magnitude – at the mouth of the Orange River. The Oranjemund deposit, running up the west coast from the mouth and now largely worked out on land, can claim a number of records, in terms of size, scale of operation and quality of stones. It was an extraordinary mine in many respects.

The geological quest keeps us on and near the beach, but forget umbrellas and sand-castles and children splashing. This is a very different kind of beach where the conditions answer a question central to understanding not only the deposit itself, but also its singularity.

The impression is of angry greyness, both overhead and the sea itself, and the wind howls. You have to shout to make yourself heard, partly because of the surf that roars far into the distance and partly because of the cobbles and boulders at your feet that clatter and crash, up and down, up and down, with the surf. As you watch the seagulls swoop out of the fog, you wonder if your memory is playing tricks on you. Wasn't the sun shining when you left the guest-house in the village? Yes, it was; the mist is coming off the sea and not reaching far inland at all.

Here the cold Benguela current that starts in the Roaring Forties has an urgency that will be dissipated as it approaches the equator. So, too, the cold polar air which sweeps the west coast with a ferocity that reminds us we're less than 700 kilometres from the tip of the continent, the closest land mass being Antarctica. It whips the sea up into a froth and sculpts beautiful crescent-shaped barchan dunes as it drives the sand off the beaches and far inland.

The question is how the diamonds brought to the sea concentrate here, and why here only. In Brazil the Jequitinhonha River, in Venezuela the Orinoco, in India the Krishna, in Russia the Lena – all carry rich endowments of diamonds along their length, but on the beaches at their mouths there are practically none.

Of the number of factors that combine to concentrate diamonds at Oranjemund, the first and foremost is the supply. The area drained by the total Orange River system includes an

Seeing the Orange River in flood, it is easy to imagine its carrying capacity.

exceptionally high percentage of cratonic crust reckoned to be the most fertile, for diamonds, on the planet. The kimberlites intruded into and through it have supplied a vast quantity of diamonds to the Orange River system. And yet, for the varied reasons of South African geological history, the dilution of the diamonds by other river-borne sediment has been less than it might have been.

Assuming there was a single major river supplying large quantities of diamonds to the so-called terminal placer[2] at the coast – like the Orange – it would still need a special combination of circumstances to gather them into a workable deposit. The desert climate of the coast, the strong longshore current and the hyperactive surf zone that you've seen at the Orange River mouth make an almost 'perfect storm' for diamond concentration.

Another important contributing factor is that the continental shelf – the shallow part of the ocean between the beach and the continental slope, where the profile plunges to abyssal depths – is wide here and 'neutrally buoyant'. It doesn't sink with the weight of sediment deposited, nor does it build a delta out to sea, as with the Ganges. The ocean floor off the mouth – and up and down the coast – has stayed at more or less the same depth under the water through periods of lower and higher sea level.

Additional attributes that favour the Orange River for concentrating diamonds at the mouth are its relatively steep gradient as it approaches the coast and the fact that most of its very extensive catchment basin is in a well-watered part of the country. Floods are not unusual, and they carry in a swirling brown torrent boulders and fallen blocks of rock from the crags through which they surge not long before they reach the mouth; these help concentrate the diamonds. Most importantly, they collect the stones themselves from terraces along the river, reaching back many hundreds of kilometres, almost to its

Merensky was the first to recognise that raised beach gravels with oyster shells carried rich concentrations of diamonds. This discovery would change his life.

multiple sources. Finally the entire load debouches at the mouth and into the ocean. There the concentrating begins, as it has for millions of years, in an environment no man could have designed better.

In 1920 there was a man who might have rued taking on beach diamonds. Sir Ernest Oppenheimer's Anglo American Corporation, by then extremely successful on the Witwatersrand, had just formed the Consolidated Diamond Mines (CDM) to reopen the Sperrgebiet, the diamond fields south of Lüderitz in South West Africa. From the beginning a combination of factors had bedevilled operations in the old German mining areas with their scattered small remnants of gravel. In 1926 the discovery of diamonds around Alexander Bay, a gull's cry south of the Orange River, alerted Oppenheimer's geologists to the possibility of there being diamonds immediately north of the river, in the Sperrgebiet but 200 kilometres south of where they had been battling the odds and the environment.

Simple, sound geology was rewarded when in their first pit the prospecting team found raised beach deposits deeply buried by wind-blown sand. Reaching them was not easy, but the gravels were there, virgin gravels, unworked by the Germans. Two years later their persistence paid off. Within the space of a few months, stones of 246 and 116 carats were found. The prospecting teams were incredulous as they pondered how many hundreds of kilometres the diamonds had travelled to get to where they were found. Transport and survival of big diamonds over prodigious distances like this are an almost unheard of phenomenon. The geologist Dr G. Beetz marvelled at the implications: mostly, he was impatient to see how far the raised beaches with their diamonds went.

Development at CDM faltered when, soon after the discovery in 1928, Wall Street crashed. The scale of work had to be drastically reduced and all but ground to a halt. But by 1934 the corner had been turned, and mining and evaluating the deposit was able to begin in earnest. CDM never looked back. It is estimated that over the 94-year life of the various operations from Lüderitz to the Orange River, between 75 and 100 million carats of diamonds were produced, 95 per cent of which were top-end gem quality.[3] The great majority have come from the CDM mine, making it one of the world's biggest diamond mines of all time.

The raised beaches that hosted the diamonds extend from the current beach-head up to an elevation about 30 metres above sea level. Near the river this terrace extends 5 kilometres inland, though as one travels northwards along the coast the bedrock rises more steeply and the strip narrows until it disappears completely, to reappear again in 'pockets' – isolated small patches of gravel in flatter coves. As you might expect, the diamonds are smaller the further north one goes.

It will come as no surprise that sedimentary processes along rivers and along the coast are different. But as far as concentrating diamonds is concerned, the same principles apply to both without a great deal of change.

A diver on his way to suction offshore diamond gravels.

Changing sea level

Before closing the section on water-borne diamond transport and concentration – and now that we're at sea level – we should reiterate a point that we've only mentioned *en passant* until now. Today's sea level, wherever you may be, is no more fixed than the needle in your car's fuel gauge. In geological time sea level rises and falls, although in the span of a human life the change may be barely measurable, even with sensitive instruments.

In quite recent geological time, sea level has fallen far below the one we know. It is all too easy to think of the current shoreline as being at the end of a downward process, from the highest, oldest terrace, inland of the shoreline, to where we know it. Not so: it's a two-way street. Sea level oscillates. Some terraces we might walk up to from the beach; others we would have to reach by donning our wetsuits and diving. Ancient terrace deposits may be raised or they may be submerged. Off the west coast of southern Africa there are still major operations under way mining submerged terraces, and there will be diamonds in bedrock potholes and gulleys deep under the churning surf for many years to come.

THE FIRST DIAMONDS to be found near Lüderitz were small – on average 10 to 15 per carat – and remarkably uniform in size. To understand why, we need to look at another agency of transport and sorting, which we've hinted at: that of wind.

We've said that stones in the hundreds of carats found at the mouth are a tribute to the extraordinary carrying capacity of Orange River floods. We've seen wide terrace gravels near the mouth getting narrower northwards and finally losing their continuity to become isolated pockets. We know that even these are increasingly rare as the mouth recedes further and further behind us. And we've seen the diamonds getting smaller northwards. The coastal sorting is far more efficient than anything a river can do with its ebb and flow. If we are looking for efficiency, there's nothing better than wind.

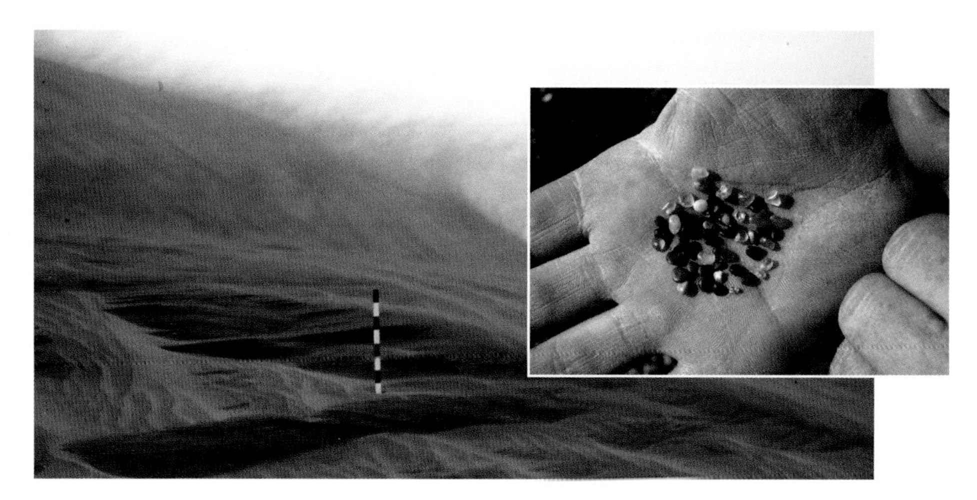

A 'smoking dune' at Bogenfels in wind gusting up to 100 kilometres an hour, with an 'unrounded' diamond in a group of grains from the desert floor.

According to Dr Ian Corbett, there are two ways in which grains of sand and grit are driven by wind, called saltation and creep. The former, which gets its name from the Latin word *saltare* – to leap – is what happens when grains become airborne, lifted by the wind and carried along, to land some distance downwind. When they land they have enough forward momentum to bump other loose grains forward before leaping again. A sand grain will move a pebble the size of an average grape, not very far at a time, but moving nonetheless. This is called creep.

Of saltation Corbett says, 'Saltation bombardment is particularly intense during gusts of 18 to 25 m/sec [25 m/sec = 90 km/h], and grain impact is excruciatingly painful on any exposed flesh. Grains impacting on hands raised 2.5 to 3 m into the air collide with incredible force, and despite wearing a ski-mask and goggles, one's face cannot withstand the pain of impacts 1.3 to 1.5 m above the ground for long. Such conditions are relatively common within this high-energy aeolian [wind] environment.'[4]

Chameis Bay, 120 kilometres north of the Orange River mouth, is about the northernmost reach of copious amounts of diamonds carried north from the mouth by longshore transport for tens of millions of years. For our story, though, Chameis Bay is not the end but the beginning. It is from marine terraces there that the diamonds Ian Corbett studied started the last leg of their journey.

The shape of the coastline gave rise to 'aeolian transport corridors', and it was in these that the barchan dunes formed and moved and where the diamonds were concentrated. Outside the corridors the wind blew – and still blows – as incessantly as in them, but there was not enough sand to get the larger grains to creep. Over millions of years of a wind regime such as this, a great deal of aeolian sorting by size, density and shape has taken place.

Corbett concludes that 'the smaller diamonds were preferentially removed by more rapid transport' while the 'coarser material [including diamonds] forming part of the dynamic creep bedload followed at a slower transport rate'. In answer to those who question whether the wind was capable of moving diamonds big enough to be of interest to miners – and there are such doubting Thomases – Corbett argues that 'only the diamonds that were too large to be transported by aeolian creep can be interpreted to be residual lag deposits resulting from deflation and aeolian removal processes'. The bigger diamonds lagged behind, the others were propelled forward. (Deflation is a mechanism of concentrating a part of the bedload by removing lighter or finer material, as in winnowing, and of giving rise to 'residual lag deposits'.)

Perhaps the reason for the scepticism that exists is that the role of creep is not fully appreciated. The biggest diamond that is likely to be moved under the right conditions by creep is about a carat. We saw in an earlier chapter that the diamonds mined by the Germans during and after 1908 – and mostly well north of Chameis Bay – were generally small, their excellent quality making up for their smallness. Not so small was the time interval over which transport by aeolian creep took place. Over tens of millions of years the finer stones were carried many kilometres by the wind, and there was enough deflation to achieve the greatest concentration ever encountered.

◆◆◆◆◆

HAVING LOOKED at one unusual moving agent of diamonds – wind – we come to the final possible transporter: ice. Let us retrace our way down the coast to near Oranjemund. Several years ago diamonds from the mining operations, both on- and offshore, were collected for a research project. Those which could be seen to contain within them tiny inclusions of a not very common mineral, clinopyroxene, were put aside until there were 52 such stones. The inclusions, none of which exceeded half a millimetre in their longest dimension and were mostly much smaller than this, were then removed from the diamonds after they had been crushed, and subjected to an age-dating technique called $^{40}Ar/^{39}Ar$ step-heating. It is a technique reckoned to provide an accurate dating of the emplacement of the kimberlite that carried the diamond (as distinct from the age of the diamonds). The project was designed to give an idea of how long the diamonds at the Orange River mouth had been present at the surface.[5]

Outwash of glacial till from the most recent glaciation.

Since the age of all the kimberlite pipes in South Africa is known, the information from the diamond inclusions provided a good indication of where the diamonds might have come from. 'Based on the ages of known southern African kimberlites, the current results suggest that very few (<11%), if any, Namibian diamonds were sourced from Dwyka glacial deposits or older sediments.' Not everyone would agree. For our purposes the main question that these two contrasting points of view poses is this: did the Dwyka ice sheets form an important 'travel agent' for diamonds? We have seen that rivers move and concentrate them, in some cases very effectively, that the 'perfect storm' off the Oranjemund coast does so even better, and that the Namib Desert wind outdoes both of them to lead, in a number of instances, to spectacular enrichments. I am inclined to agree with the view held by the age-dating specialists that the ice was not an important agent of distribution of diamonds.

Glacial tillite (300 million years old) showing its 'unsorted' nature.

A parallel debate has raged in Brazil for years, though there practically all the diamonds mined over nearly 300 years have come from river sediments with diamond-bearing conglomerate and some glacial tillite in their catchments. The continuum from diamond-bearing kimberlite pipe to nearby river shingle to downstream gravel does not exist in Brazil as it does in South Africa.

There is no doubt that the Dwyka glaciers ground over ancient kimberlite pipe-tops and gathered some diamonds. But in my view, an ice sheet laden with sediment from big blocks to fine rock flour and moving at a snail's pace is a very poor sorter of material. It could not have segregated diamonds from the rest of its load. Moreover, there is an empirical progression that is difficult to resist. Air concentrates better than clear, fast-moving sea water, which in turn sorts more efficiently than mud-laden rivers. Sediment-choked ice is surely a non-starter.

The other medium that needs to be considered is eskers. Eskers are the rivers that form under ice sheets grinding over land, and they transport ice-melt much as a river would. Geologists working on Canadian diamonds know them well: it was in esker outwash sediment that the kimberlite indicator minerals were found which led to the discovery of many of the deposits there.

In his book *Diamond: The History of a Cold-blooded Love Affair*, Matthew Hart describes the scene when Chris Jennings sent his geologist, Leni Keough, to sample for kimberlite indicator minerals. Hart writes: 'Mainly she sampled the eskers – gravel ridges formed from material washed out of melting glaciers.

'"I had a set of field sieves with me," Keough said, "and sometimes at the end of the day I'd sieve the samples down to a concentrate and look at them through a microscope. Gradually I started to find indicators. I didn't even need a microscope. They were right there – beautiful purple garnets, chrome diopsides, ilmenites."' So we know that eskers concentrate indicator minerals. But what about diamonds? Tom Nowicki, well versed in Canadian diamonds, has this to say: 'The diamonds [in eskers] must be present, but I would see this setting as being analogous to sandy river sediments with lots of indicators and a few, mostly small diamonds, unlikely to reach concentration levels that are required for economic extraction. In order to form decent alluvial deposits you need a high-energy environment with good trap sites that is operative over an extended period of geological time.'

So we see that the ice sheets themselves are incapable of doing any sorting, and their eskers are just about as ineffective, for different reasons.

I CAN'T EXPECT that you'll never cross a river again without looking to see if there's any gravel around, or see a pile of beautifully rounded river stones without wondering where the nearest craton is, or brace yourself into a howling gale without sparing a thought for Ian Corbett and saltating sand grains. I do hope, however, that you may see the world in a somewhat different light, an awesome place every part of which is a piece of a giant 4-D jigsaw puzzle, with time the most important dimension of all.

CHAPTER FOURTEEN

THE SUPPLY CHAIN

Diamond sorting at the DTC facility in Gaborone, Botswana.

The first syndicate

The route travelled by diamonds from the rock in which they're found onto a pretty female finger is more complicated than one would think. Yet it was not always so. To understand the complexities of diamond marketing today, we need to turn back the clock a little over a hundred years and see how it has evolved.

Before the late nineteenth century the process was almost as simple as with any other 'consumable' product. At the same time as new diggings proliferated in Brazil during the eighteenth and nineteenth centuries, the market for diamonds seeped downwards from the royal houses of Europe and their aristocracies to include the rapidly burgeoning new middle classes, but it was easily able to absorb the growing supply. Then came Kimberley and diamond mining on a scale without precedent. While the 'dry diggings' provided the volume of stones, the quality, with regular perfect gems in double-figure carats, continued to flow from the river diggings. *Pari passu*, the manufacturers in Antwerp, Amsterdam and London proliferated to keep pace with the supply of rough. But soon the stones flooding into jewellers' windows began to languish there, and prices fell. Investors were dismayed to see stones they had bought months previously, now offered at two-thirds of the price, half even.

When the merchants, or diamantaires, could not sell their stock, they stopped buying rough. Alarm bells rang loudly for Rhodes and Barnato in Kimberley: sliding prices in any commodity can be calamitous, but nowhere more so than in non-essential luxury goods like diamonds. The supply end of the see-saw had been heavily weighed down by burgeoning production from the Kimberley mines, leaving the demand end precariously perched in the air. Between them, the miners and the merchants had to restore the balance.

In 1890 the London Diamond Syndicate was formed as a joint initiative by the main merchants in Kimberley and London and by De Beers. As a single-channel marketing agency strategically placed at the pivot of the see-saw, the Syndicate saw it as its priority to regulate what was potentially an extremely unstable industry. The problem was the market, which has been described as 'susceptible to rumour, over-production, and organised to promote a good which was only considered to be worth buying if its price was artificially supported and raised'.[1] It was the market – the merchants or traders, who occupied the middle ground between the miners and the end-users – that sought regulation.

From the outset of the new era in diamond mining there had been close liaison between the producers – ultimately De Beers – and the main merchants in London. The earliest miners had practically no backing of their own and were furthest removed from the buyers of diamonds and their jewellery. Between them were two links in the chain: the merchants and the manufacturers. In the last two decades of the nineteenth century, the merchants, being closer to the end-market, had funds to invest and they needed merchandise. They would pay for security of supply by investing in the mines.

London's Hatton Garden about a hundred years ago: the group of men are probably diamantaires, since one appears to have a loupe to his eye.

Just as food chains in the United States pour large sums into Chilean fruit pack-sheds, the diamond merchants invested heavily in effectively the only supplier, De Beers. In the 1880s some 60 per cent of the share capital of the newly amalgamated De Beers was held by 18 diamond merchants. Not surprisingly, many of these merchants or their representatives served on the De Beers board of directors.

The cosy relationship between De Beers and the merchants was severely tested when in 1902 a Transvaal prospector named Thomas Cullinan secured the rights to a very large

Thomas Cullinan's Premier Mine, now renamed Cullinan, in the early days.

diamond-bearing kimberlite pipe near Paul Kruger's capital, Pretoria. And a short five years after Premier, as the pipe and its mine had been named, started producing in 1903, the fabulous coastal fields south of Lüderitz were found by the Germans.

A variety of transitional arrangements – none very satisfactory – were made during this time to try to contain the threatened chaos in the marketplace. Bad turned to worse with the recession of 1907–8, when the demand for diamonds all but evaporated. But in the end the evolution from a relationship between De Beers (and the smaller Orange Free State mines of Koffiefontein and Jagersfontein) and the London Diamond Syndicate to one that included the new South West African and Transvaal producers was successfully accomplished in 1914.

One who watched the goings-on with particular interest was Ernest Oppenheimer, the young German-born diamond valuer and representative in Kimberley of a London-based firm of merchants which was a member of the Syndicate. As soon as he had income at his disposal, Oppenheimer started buying shares in De Beers and in Premier, until he had large blocks in both. For a while he had to put his diamond dreams out of mind: the opportunity had arisen of staking a claim in the newly discovered Witwatersrand gold fields, with support from J.P. Morgan in New York – the origin of the Anglo American Corporation. While he set about using his finance to acquire some of the best properties on the Rand for his company, at the back of his mind were diamonds. His coup came in 1919 when word reached him that the concessions of the German Colonial Company in South West Africa, suspended during the 1914–18 hostilities, might be available.

In the aftermath of the war, the disposal of 'enemy property' was overseen by the South

The Witwatersrand mines, now the deepest in the world, start to go underground.

African government. In 1919, when the Anglo American chairman made his approach, the government gave him its full attention. Oppenheimer had planned his strategy carefully, tapping into the wisdom of his older brother, Louis, in London throughout his negotiations by letter and by telegram. In the end his persistence secured transfer of the rights to a huge strip of land between Lüderitz and the Orange River to Anglo American. It would turn out to be one of the most important single acquisitions in world mining history.

De Beers was appalled. When given the news of the Anglo American acquisition, Ernest, writing to Louis, related that 'Sir David Harris [a senior De Beers director] ... was beside himself with rage and threatened all sorts of dire punishment'. Other directors, recognising there was no stopping Oppenheimer, took the news more philosophically. Hedley Chilvers, in *The Story of De Beers*, a classic of diamond historical writing, accords the event, pivotal in global diamond history, one understated sentence.

Oppenheimer had shown the world that he was ready to take over Rhodes's mantle as the king of diamonds. He was well connected in the Syndicate, owned large blocks of shares in both De Beers and Premier, and his gold company had secured exclusive rights to a large, still relatively unknown tract of diamond-bearing land in South West Africa. All the diamond rights in the only important producing corner of the world were soon to be consolidated under one umbrella.

The peace that settled after the end of the war in 1918 was followed by devastating chaos in markets and currencies. Severe depression swept through the world in 1920 and 1921, exacerbated, for De Beers and diamond merchants, by the offloading onto the market by the

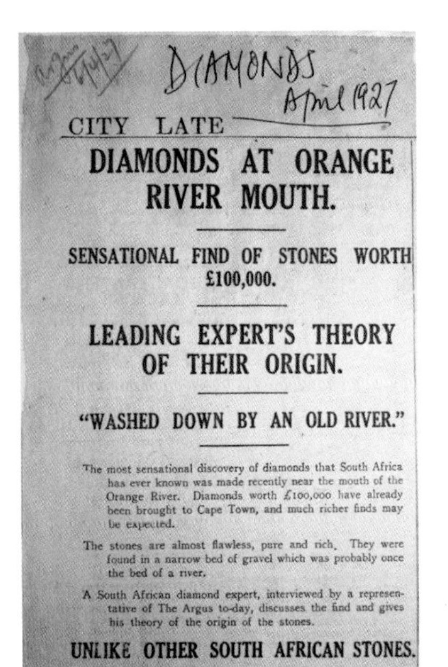
DIAMONDS
April 1927

CITY LATE

DIAMONDS AT ORANGE RIVER MOUTH.

SENSATIONAL FIND OF STONES WORTH £100,000.

LEADING EXPERT'S THEORY OF THEIR ORIGIN.

"WASHED DOWN BY AN OLD RIVER."

The most sensational discovery of diamonds that South Africa has ever known was made recently near the mouth of the Orange River. Diamonds worth £100,000 have already been brought to Cape Town, and much richer finds may be expected.

The stones are almost flawless, pure and rich. They were found in a narrow bed of gravel which was probably once the bed of a river.

A South African diamond expert, interviewed by a representative of The Argus to-day, discusses the find and gives his theory of the origin of the stones.

UNLIKE OTHER SOUTH AFRICAN STONES.

Bolshevik government, whose rouble was in tatters, of large quantities of beautiful gems at knock-down prices. American offtake dropped to a third of its previous levels and De Beers was forced to scale down operations drastically and to stop mining and washing altogether.

These circumstances forced substantial change on the Syndicate. The De Beers history records that the company accepted an offer from 'Sir Ernest Oppenheimer and friends', later referred to as 'the new (Oppenheimer) [*sic*] Syndicate', to buy all the company's production for five years from the beginning of 1926. In April that year Oppenheimer was appointed a director of De Beers, filling a vacancy resulting from another board member's resignation.

The timing was fortunate. An article in the *Cape Argus* dated 6 April 1927 announced: 'The most sensational discovery of diamonds that South Africa has ever known was made recently near the mouth of the Orange River. Diamonds worth £100,000 have already been brought to Cape Town, and much richer finds may be expected.' Sir Ernest Oppenheimer had not known what to expect when he was invited by the government to see the first parcel of Namaqualand stones brought to Cape Town. He could scarcely believe what he saw. As he regained his composure he realised that, as a short-term measure, not only did his Syndicate have to buy those stones and the full interim production from the field, but, in the longer term, he had to gain control of it. By 1929 he had secured control of most of the new field and could ensure that the Namaqualand production was routed to the market via the Syndicate, of which he was firmly in control. Later that year, three and a half years after he had been appointed to the De Beers board, he was made company chairman.

Oppenheimer would find that his firm hand on the De Beers tiller could never be lightened. America and Europe entered a depression without precedent following Black Thursday in October 1929 on the New York Stock Exchange. Diamonds were already coming out of Angola, the Gold Coast (present-day Ghana) and the Congo in large quantities; in 1929 they were discovered in the Ivory Coast and a year later rich deposits of top-quality stones were found in Sierra Leone. In 1931 the first discoveries were made in Central African Republic and five years later Guinea became a producer. Compared to the De Beers stream, they were little more than a trickle, but Oppenheimer monitored all the developments.

Soon after the Namaqualand discoveries were made, Oppenheimer got the break he needed when just north of the Orange River, in his Consolidated Diamond Mines concession, a trove of top-quality stones beyond even the dreams of his German forebears was found

NICKY OPPENHEIMER

Grandson of Ernest, son of Harry, Nicky Oppenheimer had big shoes to fill. He has done so admirably, his legacy comprising three singular contributions to the industry and South Africa. He has been conspicuous at the forefront of the curtailment of the trade in 'conflict diamonds', with the main measure introduced in this quest being the Kimberley Process Certification Scheme (KPCS). Then, as De Beers navigates increasingly choppy marketing waters, Nicky's hand is firmly on the tiller with its 'Supplier of Choice' strategy, to 'drive consumer demand for diamond jewellery in the context of an increasingly competitive luxury goods sector'. Under Nicky, De Beers has pursued long-term growth at the retail level, at the same time making a real contribution in the countries where it mines (particularly Botswana) through its commitment to local beneficiation. Finally, Nicky and his wife Strilli have dedicated themselves to environmental conservation projects across Southern Africa, restoring the natural environment to its original state. The Oppenheimers care passionately for the world around them.

below the shifting sand dunes. The deposit would sustain a mammoth mining operation for decades. Oppenheimer's De Beers Consolidated Mines, together with Anglo American and their subsidiary companies, now stood squarely astride the world diamond industry.

Having served its purpose with due effect, the Syndicate finally gave way to the Central Selling Organisation (CSO), part of De Beers' London-based Diamond Trading Company, formed in 1934. The next thirty years saw more than 90 per cent of (formal) global production of a steadily growing stream of rough diamonds sold through the CSO.

For as long as De Beers produced nine out of every ten diamonds legally tendered from its massive South African mines, it released them into the market through the CSO. The size of the stockpile held in its vaults was a matter of speculation to anyone outside De Beers and most within. Merchants had access to enough stones to sell on to manufacturers, who, in turn, could keep their market supplied. In this relationship the CSO could offer the necessary security to keep the mines producing profitably and the owners of high-price diamonds were assured that their jewellery was always worth more than when they'd bought it.

The supply pipeline in a shrinking world

Diamonds are the only billion-dollar industry where the product consists of individual items each of which is unique. Other major mined commodities – gold, silver, copper – are sold by weight while, at the bottom end of the scale, individually valued stones besides diamonds, whether magnificent emeralds or sand-size garnets, do not exactly constitute a massive industry.

A collection of diamonds from Gem Diamonds' Cempaka operation in Kalimantan.

Diamonds as they come out of the ground and into the sort-house vary in a number of respects: mainly size, quality, clarity and colour. Different mines, sometimes even different parts of the same mine, have different proportions of gems and industrials, big stones and small stones, crystalline stones and fragments, clear stones and cloudy stones, perfect white, off-white and 'fancy' coloured stones. At some point, before being offered to the trade, they have to be sorted and valued.

Every day masses of stones come out of the ground, mostly worth a few dollars a carat, some tens of thousands, and everything between. Every day artisans set hundreds of tiny diamonds, unrecognisable as such, in saws and drill bits, while far away a master craftsman studies a single gem the size of his thumb tip, slowly coming to terms with it, reading its internal architecture, realising how he must cut and facet it. Between the artisans and the craftsman there is a vast range of different skills. How can such a complex equation be balanced? Yet it is, day after day, thanks very largely to the Diamond Trading Company and the remarkable single-channel pipeline that has been developed over the decades. Though some give it grudgingly, the homage is due.

The simple part of the distribution process involves industrial diamonds. They can be made in factories and the price accurately established. Mine production of industrial stones competes with them. There is a huge demand and a carefully balanced supply, so they can be traded on the open market with no danger of chaos in the pricing.

A rung above them in the hierarchy of quality is occupied by so-called near-gems. These are borderline natural stones that are generally good enough to be faceted for the mass market if it will take them, failing which they will wind up on an industrial tool. They, too, present little threat to a stable market. Like the bottom-end industrials, they are not what the miners are digging for.

The Diamond Trading Company's vital role lies in the gems, where it bridges the space between what the miner has to sell and what the customer is looking for, either in Tiffany's for a very special occasion or in the local supermarket on the spur of the moment. At present the company has ten sales or 'sights' of rough diamonds a year, to which 75 to 100 sight-holders are invited. They are held simultaneously every five weeks or so in four centres, London, Kimberley, Gaborone and Windhoek, and last from Monday to Thursday. Sight-holders are carefully and strictly vetted by the company for their financial, technical and ethical profiles.

Before each sale sight-holders submit a breakdown of the sort of material they are looking for. The 'sight' or parcel of goods they are offered at the sale should include most of what they have asked for, plus a lot else besides. They are told what the price is to be for the sight, a topic on which there is no negotiation. They will register no surprise that it is well over a million dollars. Then they sit down to study the goods they have been offered before agreeing to the purchase. And they *will* agree if they wish to retain sight-holder status.

We come now to an aspect that has characterised the diamond business from the start. Most of it is conducted on the basis of trust: the value of a parcel of diamonds, a handshake, a man's word. Not only has the Diamond Trading Company – and before it the CSO – been doing this for many decades, but it has a massive investment in ensuring its continuance. Today De Beers is pouring fortunes into new mines, new developments in its existing mines, new technology and a public relations enterprise few companies can dream of. Far more than any sight-holder, it needs to see a stable industry, with price levels sustained and preferably gradually increased.

DTC's most modern complex, in Gaborone.

Let's create an imaginary scenario. The sight-holder, a merchant, leaves Charterhouse Street in London for his business in Antwerp. He has three manufacturers who he knows will take half of the box at his side. The stones he does not already have orders for he will take to the Antwerp Diamond Bourse, which was the first trading exchange to deal exclusively with rough diamonds. Like the other main exchange in Antwerp, the Diamantklub van Antwerpen, it lies off Pelikaanstraat, which forms the eastern boundary of the *diamantsektor*. This is to the downstream end of diamonds what Kimberley and its Big Hole are to the mining end: the centre of gravity.

The diamond district accommodates around 4000 diamond manufacturers and traders, and anything between 15 and 20 billion dollars pass through the bourses every year. Its three or four streets are a buzz of activity, come rain or shine. Our merchant will call on a few manufacturers in the bourse; probably last, but by no means least, on Rosy Blue, where he will hope to find one of the Mehta family in attendance. They will almost certainly take his smaller stones off him, since it is practically only the Indian manufacturers like the Mehtas who can cut a gem of 0.005 carats, or a thousandth of a gram, smaller than a pinhead.

If Rosy Blue does buy his small stones, they'll send them to Surat, in northwestern India, for cutting. There the Mehtas outsource some of their cutting to families who put their children – whose eyes are so much sharper than any adult's – to work. At the other end of the scale they have highly mechanised polishing works, using the most modern technology. Surat and other provincial centres and villages produce many millions of tiny polished diamonds every year, most of which find their way to the United States, where these days even university students buy their girlfriends diamond jewellery.

Polishing diamonds in India.

Every diamond faceted into a gem has had eyes study it for hours: bright, young, near-black eyes of Indian youngsters, or watery, faded blue eyes of grizzled third- or fourth-generation Belgian craftsmen and, before that, eyes of African and British sorters.

Cutting

In a parcel of stones, what do we see? In a perfect world, with infinite time and no directional stress, a diamond will grow as an eight-sided octahedron, two four-sided pyramids joined at the base. In some kimberlites we find lots of these, mostly small and breathtakingly perfect. In the trade they are called stones (although miners, and we in this book, usually use 'stone' and 'diamond' synonymously). We also see bigger crystals, which are commonly resorbed, giving them a rounded look, though you can sometimes still make out their crystal form. The sorters and manufacturers call these shapes. Some deposits, like Letšeng in Lesotho, produce very few *stones* or shapes, but mainly biggish, sometimes very big, cleavages, which are irregularly shaped and appear to have been broken, perhaps during a late stage of their ascent from the depths. Macles (or maccles) are flattened triangular crystals whose shape is due to the fact that they are twinned. Lastly, the trade recognises flats, of irregular shape, thinner than macles and of no particular outline.

To understand how the manufacturers set about their work, look at the diamond in your or your wife's ring. In essence it consists of three parts. The top part is called the crown, which is mostly what you see; between that and the deeper, lower part below the clasps, which is called the pavilion, is the girdle, a narrow 'belt' around the middle.

As from the earliest days good octahedral diamonds were found, with well-defined (flat) faces, (linear) angles and (point) corners, it was inevitable that early cutting would seek to accommodate this geometry. Initially, shaping the hardest material known to man meant that new styles of cutting evolved extremely slowly. Gradually, as more tests were carried out, cutting and polishing techniques improved. Then, during the Renaissance, science was brought to bear and instruments were developed to measure properties of materials, like lustre, refractive index and so on.

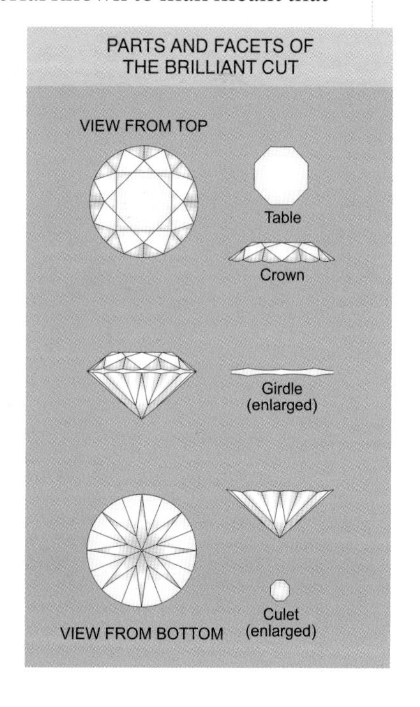

Lustre is the measure of light reflected off a polished surface: diamond reflects 17 per cent of light falling on a perfectly polished surface, compared to 5 per cent for glass. The life of a stone combines the lustre and the internal reflection, which is the measure of how much of the light not reflected off the outside surface is reflected back off the inside of the pavilion facets. Since this depends, firstly, on the angles between the crown and the pavilion and girdle between them and, secondly, on the angles of all the facets in the crown and the girdle to each other, the life of a cut diamond depends largely on its cut.

As distinct from the life, the fire in a diamond results from its refractive index. This describes the effect of seeing different colours reflected back at you out of the stone. This is no different from the way that sunlight passing through a shower of rain breaks into the colours of a rainbow as the white light is refracted by the droplets of water. And if water (refractive index or RI = 1.33) refracts white light well enough to dazzle us from time to time in rainbows, diamond, with an RI of 2.417, does so much better; much better, too, than quartz (RI = 1.54) or topaz (RI = 1.63) or any other 'white' stone.

But what has interested manufacturers and technicians in the diamond business more than the fire – which just confirms they have a superior stone – is the life. The stone has to be cut to maximise the internal reflection. This became the Holy Grail. In 1914 Marcel Tolkowsky, a Belgian engineer from a diamond-cutting family, and a cousin of one of the great cutters of all time, Lazare Kaplan, produced a theoretical treatise giving the dimensions that would achieve this objective. His book on the subject, *Diamond Design*, was published in 1919, and his Brilliant Cut remains by far the most common cut in use today.

The prescribed angles, and consequently the relative depth of crown and pavilion, are well known to all serious manufacturers and adhered to by anyone with a reputation to protect. Inevitably, though, there are those who cheat. They sell a gem that looks to all intents and purposes like a Brilliant Cut, but a really good look will show that it lacks the life of a true Brilliant Cut. The reason is to reduce the loss of mass during cutting, in other words to get a better return on the outlay in the stone. A Brilliant Cut diamond of 1.05 carats means starting with a rough stone of a certain weight; depending on its shape, this may be close to 2 carats. And such a stone may cost hundreds of dollars more than one of 1.7 carats of the same shape, from which, by altering the proportions, the unscrupulous manufacturer will still be able to cut a 1.05-carat gem.

Brilliant Cut 'champagne' diamonds from Rio Tinto's Argyle Mine.

This brings us to the aspect of weight loss or waste. It goes without saying that a lot of the diamond that comes out of the ground is lost before it's ready to mount in a ring. This is especially so with big stones that can be cut or cleaved so as to yield a number of gems. For example, the Cullinan, which weighed 3106 carats when it was found, was cut into nine gems with a total weight of 1055.9 carats, meaning that 65.25 per cent of the original stone was lost. More recently the Lesotho Promise, from the Letšeng Mine, was cut into 31 gems with a total weight of 223.25 carats. Nearly 400 carats (63 per cent) of diamond was washed down the drains of Antwerp. In the value chain it matters not: the sale price of a large gem, and particularly a string of large gems from a single rough mega-diamond, is so astronomically high that it more than compensates for the loss.

It is a matter of simple economics for manufacturers to make sure that they get the most 'gem from their rough', without sacrificing life. This means they may decide to cut the stone but leave some small inclusions in the cut gem, knowing that this will downgrade its value. It depends on whether the trade-off for the lower quality rating for the bigger size makes it worth while. In fact, to see an inclusion, or several, would not be unusual: most rough diamonds have them. Consequently diamonds are sorted, or graded, according to the offensiveness of the inclusions, or their clarity. They are termed, according to their clarity, Flawless, IF (internally flawless), VVS (very very small inclusions), VS, SI and so on.

One choice which used to be as challenging was whether to saw 'stones' and 'shapes' (described as 'sawables') for two gems or to get a single, probably slightly bigger, gem out of the rough. Today the decision is made easier by technology developed by the Israelis. This offers 'an automated computerised product for assessing a diamond's proportion and symmetry, key parameters in the grading of a diamond's cut'. Stones not sawable are described in the trade as 'makeable'.

Though normally 'white', diamonds may show a range of colours.

The sorter will have to decide what colour each diamond is. At present the commonest system of colour grading is that set by the Gemological Institute of America, in which a letter of the alphabet denotes the colour. The best is a 'D'-colour stone, corresponding to Exceptional White + of the World Jewellery Confederation. Then come 'E' (Exceptional White), 'F' (Rare White) and so on, down to 'L' (Tinted White), with 'M' to 'R' described as Tinted Colour. Needless to say, the value decreases as one progresses down the alphabet, and you just have to believe that the experts, with highly trained eyes, can see the difference. To an extent, it's a matter of perception and taste, so that a low-value brown stone may jump a few rungs up the value ladder if its seller can persuade you it is not brown but 'cognac'. Beauty, after all, is in the eye of the beholder. Intensely coloured stones, the 'fancies', are highly prized and valued, especially the pinks and blues. These are just some of the dilemmas facing manufacturers and, to an extent, the sorters, who start the process which ends at the diamond polisher's bench. One will look at the model – is it a stone, a shape, sawable or makeable? – while another will look at the clarity and yet a third at the colour.

By far the biggest sorting operation in the world happens in the Diamond Trading Company. Here they start with diamonds that have been graded according to size, using a tried and tested system of sieve-like screens, so that the size groupings reflect very closely the weight of the stones. (Weight, measured in carats (1 carat = 0.2 grams), is the criterion as far as jewellers and 'consumers' are concerned, not size.) The sorters work in rooms that run the length of the buildings; in London the rooms face north, in southern Africa south. They have been designed so that the light from outside is maximised by the windows being tilted upwards towards the sky at a carefully calculated angle. Because of their facing, no ray of direct sunlight enters the room. Shadowless, the illumination is even and constant.

Not only does the room run the length of the building, so too does the one continuous desk directly below the windows, each sorter's part of the desk corresponding to an entire single-pane window. A sorter is well aware of his or her neighbours on both sides, and there is an air of studied concentration, the silence broken only by the low murmur of voices as sorters voice to themselves their doubts, questions and decisions. There are regular breaks for tea or coffee or just for leg-stretching. Each sorter works on a clean white sheet of paper and uses a pair of magnifying goggles, which he wears all the time. In his hand is a flat spatula for pushing stones to one heap or another, on the paper in front of him a pair of tweezers and a loupe, magnifying to 10 or 20 times, for problematic stones that need special attention. From the heap of hundreds of diamonds spread out in front of him at the beginning of the day he looks at each and every stone before pushing it into the heap where it belongs, the finest crystals at the top left, the poorest and smallest at the bottom right.

Sorting diamonds at the DTC in London.

And so it goes hour after hour, day after day. There is constant monitoring of the less experienced sorters by those who have been at it for years and for whom the decision of where to assign any stone is second nature. Quality control supervisors travel regularly from one sorting facility to another to ensure that not only does the quality of work pass muster but that there is consistency from one office to another.

This is the world according to the Diamond Trading Company, which sets the standard and can never stop in its quest for perfection. At the other end of the scale are thousands of humbler places where 'rough' is graded and valued, from refurbished containers on the African veld to back rooms in houses. From there the diamonds go to tender houses for the monthly tender or to exchanges around the world, big and small. From here they pass to the manufacturer, on to the jeweller's work bench, into his window and onto a finger.

In the middle of the chain are highly sophisticated laboratory-like rooms and complexes in high-rise buildings in Canada, Russia, India, Australia and South Africa. These are relatively new. Fifty years ago the CSO controlled practically all the formal marketing of diamonds through an almost untraceable undercover, maze-like channel. The politics were all wrong. But as production behind the Iron Curtain scaled up, the Soviets started marketing their rough directly in Antwerp and elsewhere, and moved on before long to cutting and polishing their gems at home, even if the jewellery market in the Communist world was small, in the early days at least. Moscow was careful, though, to maintain an association with the CSO that, if not always cordial, at least offered access to the main stream of diamonds. The Australians, too, started with the Diamond Trading Company, but because of the preponderance of tiny gems and near-gems from Argyle, the growing relationship with Indian manufacturers – for the American mass market – soon rendered the company irrelevant. And joint ventures in Canada led by BHP and Rio Tinto chose to market most of their rough independently of the Diamond Trading Company. The face of the global diamond industry has changed drastically in fifty years.

It's an amazing concept, if you stop to think about it: the network of routes around the globe travelled by millions of stones mostly no bigger than a carob seed or a small green pea.

The underworld

This closing chapter sets out to track the diamond pipeline from the various points of origin to the shop window and into the future. While we have touched briefly on the visible routes, we need to remember that there are thousands of diamonds that travel long circuitous paths under cover. Billions of dollars' worth of diamonds are traded illicitly every year. Most come from Africa, with Brazil and Venezuela trailing quite far behind. In times when political correctness is observed in most countries, even the

A scene from the film Blood Diamond.

•Hijack hero's family sadness ...Page 3

Sunday Times

MAY 28, 1972. No. 3447. Established 1906. Price 15c

THE PAPER FOR THE PEOPLE

BULLET-RIDDLED 727

Dawn drama at Chileka

BULLET HOLES, some ringed, scar the sleek body of the hijacked SAA Boeing 727 jet. This was the scene — pictured by the [illegible] of Sunday Times photographer DEREK MAY — at Chileka Airport, Blantyre, after troops launched a dawn attack on the beleaguered hijackers. A heavy calibre machine gun opened fire. Bullets punched their way through the jet. Minutes later the hijackers, hands above their heads [illegible] selves up. It was all over in 20 minutes. The drama that began at 2.01 pm on Wednesday in the sky above the Transvaal ended on this dusty airstrip in Malawi. Now the epilogue will be played out in the courts of Malawi and South Africa.

Fouad Kamil's failed hijack attempt reached the front page of the Sunday Times.

informal trade is mostly free of the taint of conflict diamonds, as Kimberley Process Certificates will attest.

In May 2000 southern African nations met in Kimberley. The gathering followed United Nations action against Unita in Angola four years earlier, specifically targeting the diamond trade involving the rebel group. If the trade, amounting to massive numbers, could be shut down, it would effectively close the tap in the arms supply and ultimately the guerilla campaign would dry up. That, at least, was the argument, and it was considered worth pursuing. A follow-up meeting in Pretoria led to the formulation of the Kimberley Process Certification Scheme, which was adopted by the UN General Assembly in December 2000. Countries where diamonds were not being used to fund armed rebellion would be duly certified and merchants were expected to buy diamonds only from certified suppliers.

If the Angolan civil war was bloody, it was nothing like that waged in Sierra Leone, where Leonardo DiCaprio's Blood Diamond was unearthed. There the Revolutionary United Front, aided and abetted by neighbouring Liberia's president Charles Taylor, fought a much more brutal war for eleven years, with large-scale involvement of women and children and horrendous atrocities. For outright lawlessness, in diamonds at least, Sierra Leone probably takes the cake. No clearer picture of the trafficking of diamonds from diggings in that country and out to Beirut via Monrovia in Liberia could be given than by 'Flash Fred' Fouad Kamil. Kamil was so successful in temporarily all but halting this trade that he drew the attention of De Beers, which decided to use his skills and less than orthodox techniques for its own ends. For a while all went swimmingly. Part of the strategy he used to penetrate the diamond underworld in quest of the original source of 'leakage' in various De Beers operations in South Africa and South West Africa was to buy diamonds himself illicitly. After a period of thorough exploration of the various networks, he became convinced that De Beers operatives themselves, up to the highest level, were complicit in the illicit diamond buying (IDB) or smuggling rampant in the world's most diamond-rich quarter.

When he tried to bring this activity, and a query regarding payment he claimed he was owed, to the attention of the chairman of both Anglo American and De Beers at the time,

Harry Oppenheimer, he found himself against a brick wall. In his view, either the chairman's mail was being censored or he saw fit not to reply to Kamil's letters. The situation had become desperate: desperate measures were necessary.

Kamil and an accomplice, Ajej Yaghi, a Lebanese inspector of police, hijacked a South African Airways Boeing 727 from Salisbury (now Harare), as it was about to commence its descent into Johannesburg. Although two ANC operatives in Harare, whom Kamil had run into fortuitously and who with their own political motives had inspired the hijack plot, failed to check in to the flight, the two Lebanese went ahead with their plan anyway. After returning to Salisbury to refuel and proceeding on to Blantyre, the skeleton crew and five hostages whom the hijackers had held on the plane endured a harrowing twelve hours on the ground there until Captain Blake Flemington cleverly and courageously effected their escape. It was, however, only the intervention of the Malawian armed forces with automatic rifles that persuaded the hijackers that their game was up and they handed themselves over, Yaghi limping.

He had taken a bullet, though it was miraculous that either escaped with his life. Blake Flemington, in an account he was good enough to send me, describes this part of the drama: 'The following day, Friday, the Malawian President Kamuzu Hastings Banda gave the order to end the hold-out forthwith as he had made arrangements to fly to Britain on British Airways and the airport had therefore to be operational. This was conveyed to the hijackers, who asked for a priest to be called to administer to them the Last Rites. This was done and the soldiers of the Malawi army lined up alongside the B727 and opened fire with their automatic weapons. Although over a thousand rounds were fired, there were only 27 bullet holes in the aircraft. At the first pause in the firing, the hijackers came running out with their hands raised. Yaghi had ironically been shot in the foot.'

Unwilling to be extradited to South Africa, where he would have IDB added to his charge sheet, Kamil persuaded President Hastings Banda to arrange for them to be tried in Malawi. After spending a few years under house arrest in Blantyre, Kamil was released and moved to Spain, before settling on the coast of Brazil, with the wide Atlantic between him and Mr Oppenheimer.

As he watches the sunrise, I wonder if, in his mind's eye, Fouad Kamil, now in his mid-80s, sees over the blue horizon to Oranjemund on the Namibian coast. During his time of sleuthing for Anglo and De Beers he would have visited the Consolidated Diamond Mines (CDM) operation there from time to time. It was a massive venture in more ways than one: not only did the mine cover hundreds of square kilometres but at the CSO sights stones from CDM made up a sizeable proportion of the offerings. It was also known to be a major source of IDB material.

One notable attempt at smuggling stones out of the Sperrgebiet, as the CDM property was known, was documented by Ian Fleming in *The Diamond Smugglers*, the only non-fiction book that the creator of the legendary James Bond could not resist writing. Far more irresistible were the diamonds Fleming's protagonist saw as he sampled a prospecting trench far to the north of the area being mined at the time. He sorted the concentrate from the little prospecting plant at the head of the trench on his own: no one would miss one or two of the beauties. Every time he sorted he picked out a few and dropped them into a jam jar he kept hidden in a corner of the sorting room, no more than a prefabricated corrugated-iron shack. Once a week the mine geologist called on him and collected the logs he had made of the trench and the diamonds from the most recent logging. It was a rich trench and the geologist was well pleased.

The Auster Autocrat with the crumpled wheel.

His resignation in November 1952 was regretfully accepted. He had been thorough and quite accepting of the lonely conditions. Perhaps there were some at his farewell party who joked, 'Come and pay us a visit some time.' They little expected ever to see him again, so they would have been surprised when it was reported that two CDM officials had brought him and a pilot into Oranjemund for questioning just before Christmas. It was not long before the truth came out.

They had landed a light aircraft on the beach near the prospecting trench a few nights before – it had been full moon and the beach was hard-packed at spring low-tide – their mission to collect the jam tin of diamonds buried at the head of the beach. They had crashed on take-off. They found themselves almost literally between the devil and the deep blue sea: raging Atlantic on one side, waterless desert on the other and the tightest security imaginable at both ends. A sentence of nine months' hard labour for the 2276 carats in the jam tin, worth today probably over half a million dollars, cannot be considered unduly harsh.

Working in the northern diamond area of CDM as a student in 1964, I drove with the group geophysicist to Kolmanskop to collect a part for one of his instruments, which had been flown from Johannesburg to Lüderitz and driven the 30 kilometres to the security gate at Kolmanskop. On the way to the gate Ken drove up to a very large shed, the biggest building in Kolmanskop. He stopped and got out. 'Come and have a look at this,' he said.

With difficulty he pulled one of the creaking rusted doors open. As my eyes adjusted to the gloom from the harsh desert glare, I saw suspended in the rafters a light aeroplane. Ken smiled as he turned to me and we walked back to the Land Rover. No words were necessary: he knew I had read Fleming's account. There it was: the ill-fated Auster Autocrat.

Back at university on the other side of South Africa I mentioned having seen the plane to a lecturer who I knew had worked at Oranjemund. He chuckled. 'Ja, I knew the guy quite well,' he said. 'Hell of a nice guy. I worked in the adjoining area. Because he was only a prospector and I was a geologist, I had a Land Rover but he didn't. One Saturday a month he'd walk over to my camp, we'd play chess and have supper together, and then I'd drive him back to his.'

And that, I'm afraid, is as close as I've been to IDB. Having watched much more recently diamonds tumble from a sorting machine on to a conveyor belt half a metre in front of me and felt that visceral stirring, I cannot help wondering what I would have done.

A last word on blood diamonds. Like Osama bin Laden, they're not going to go away: they are just too easy to get hold of, to move and to get rid of, unlike Osama bin Laden. I have a 2003 BBC report in front of me that reads: 'At the time of the 1998 bombings of the United States embassies in Kenya and Tanzania, al-Qaeda allegedly transferred cash into high-value commodities, including diamonds.

The aftermath of the 1998 al-Qaeda bomb blast in Nairobi, using diamond-laundered funds

'Several members of al-Qaeda's inner circle bought gems in Liberia and from Revolutionary United Front (RUF) rebels in Sierra Leone, according to research first published by the *Washington Post*.'

And a few paragraphs down: '*Global Witness* has estimated that al-Qaeda laundered $20m through purchasing diamonds.'

There is no reason to suppose that the dirty trade has stopped.

The case for synthetic diamonds

With all the trouble about conflict diamonds and IDB and smuggling, and, on the other hand, with the technology developed to manufacture in the laboratory not only industrial diamonds but gems as well, and more cheaply than we can mine them, why do we still dig them out of the ground? Because, say diamond miners, they are the real thing. Most of us are romantics at heart, and it's to that fundamental human trait that natural diamonds appeal.

Nothing man-made can replicate the miracles of nature that give us earthly diamonds. Not even Albert Einstein could match the magic of their creation billions of years ago and hundreds of kilometres underground. No white-coated technician can reproduce their transport from the depths to the surface with such speed that they don't oxidise or revert to graphite. Nor can high-pressure presses tell of their wanderings over the surface of our planet, moved by creeping ice sheets, raging floodwaters, pounding breakers and the stormy blast.

All these barely comprehensible wonders are encapsulated in a dazzling star the size of a carob seed, dug out of the hard Earth by monstrous machines, and carried with hundreds of tonnes of jagged rock hither and thither across a mine property, by truck and conveyor belt, before the first eyes see it. On it goes, under armed guard, to merchants on the other side of the world, to a fifth-generation craftsman off a road named Pelikaanstraat and on to a jeweller in Bond Street. People readily pay more for miracles. There are, however, those who, playing down the romantic appeal of earthly diamonds, focus attention on the inhumanity and disregard for the environment so manifestly portrayed in *Blood Diamond*.

Tiny synthetic diamonds.

Countering this accusation, companies today go out of their way to comply with every requirement demanded by the environmental lobby. Someone has gone as far as to describe the mining in the Canadian Arctic as 'an environmental project with some diamond mining on the side'. Elsewhere, at least in jurisdictions where transparency is not only accepted but expected, measures are required to ensure that rehabilitation of the mined area will follow mining. Hefty bank guarantees must be lodged, even before exploration can start, to ensure that, however profitable the mining may or may not turn out to be, restoration to pre-mining conditions is underwritten, even if it has to be undertaken by a third party.

Considering the benefits that mining, and the secondary industries that usually follow it, can bring to needy communities, it is difficult to argue that all mining is ecologically harmful. Look at Botswana. We should not for a moment turn a blind eye to the deleterious effects that the displacement of a San community is having in one particular area. But the advent of diamond mines in that country in the 1970s and 1980s saw it rapidly become a model of economic development that has made it the envy of the rest of Africa. Apart from the San issue, there are those who question the ethics of the deal between De Beers and the Botswana government, saying it was strongly biased in favour of the mining company. That is a question too complex to be debated here. Instead let us close by saying that diamond mining has led most Batswana to prosperity that would have been undreamt of fifty years ago.

There remains, nonetheless, a clear imperative for miners to be accountable and to ensure that their hands stay clean. For as long as they can do that, and for as long as sellers of synthetic diamonds certify them clearly as such, the two industries can survive peaceably side by side. It's a free world, and if cost is the overriding consideration, let cheaper synthetic diamonds be sold.

The future

During the calamitous year and a half, from mid-2008 to the end of 2009, when more than half of the world's big mines went on standby for months on end and no one bought luxuries they could ill afford, many asked themselves whether the industry would survive the crash. It is too soon to be sure (early February 2010), but the signs seem to be saying that the diamond business is alive and well. The great unknown is China. On 24 January 2010 the official Chinese news agency Xinhua reported that diamond imports through the Shanghai Diamond Exchange had risen to more than $1.5 billion in 2009. It may be small fry compared with the $12 billion that the United States bought in 2009, but for the first time it far outstripped the $600-odd million imported by Japan.

De Beers retailing to the Chinese market.

Even if the taste which the burgeoning Chinese middle class has developed for Western brand names is general, their adoption of the hitherto alien concept of diamonds in rings and other jewellery owes a lot to the low-key but relentless De Beers campaign of raising the awareness of diamonds in the newly affluent Middle and Far East. So, whether they like it or not, the lesser diamond miners have a lot to thank De Beers for: a stable industry after a hundred years of nay-saying by wise men, and a future full of hope on a playing field that is every year more level.

The International Diamond Exchange report on cut diamond prices in January 2010 closes quite lyrically. While diamond price movement is not as predictable, it notes, it is at least as inevitable as 'the rising and falling ocean tides'. Here's the last word of the commentary, unequivocally sanguine: 'Demand for diamonds and diamond jewelry will recover.'

Sorting a hundred years ago was different from today, though not altogether so.

GLOSSARY

aggradation The process of building up by deposition of sediment
alluvial Describes recent unconsolidated sediment
anomaly An unusually high or low value in a geochemical or geophysical sample
artisanal Describes the workings by artisans or small-scale miners
asthenosphere The non-rigid part of Earth's mantle below the rigid lithosphere
barchan dune Mobile crescent-shaped dune, concave on its leeward side
basalt Volcanic rock formed from basic (non-acidic) lava, usually grey or mauve
berm Sand wall constructed to keep the sea from flooding the land
calcrete Hard surface crust consisting mainly of calcium carbonate (lime), formed in arid climates
carat The unit used for weighing diamonds, equal to 200 mg
carbonado A type of diamond, usually black, comprising a tightly bound aggregate of minute crystals of diamond
Carboniferous Period The geological time interval from approximately 345 to 280 million years before the present
chromite A mineral of the spinel group consisting mainly of chromium oxide
colluvial Describes a mixture of alluvial and locally derived rock fragments
concentrate (mining) The fraction of a sample or mine feed naturally or artificially concentrated by virtue of its being heavier than (or different in any other way from) the sample or 'run of mine' plant feed
conglomerate (geological) Sedimentary rock consisting mainly of pebbles, cobbles and boulders, usually a combination of all three
craton The core of an ancient continent that has been stable for billions of years
Cretaceous The geological time interval from approximately 140 to 65 million years before the present
degradation The destruction of rock or sediment by erosion and transport, usually by water
diamantaire Usually used for a skilled craftsman who cuts and polishes rough diamonds to produce a gem ready for setting
diopside Rock-forming mineral of the pyroxene family, the chrome-rich variety found in some kimberlite and in lherzolite mantle xenoliths
eclogite In kimberlites a mantle-derived rock consisting mainly of omphacite pyroxene and garnet, commonly derived by ocean-floor basalt forced into the mantle during plate tectonics
esker A ridge marking a channel formed in a decaying ice sheet
facies Part of a rock body distinct from other parts by its composition and genesis
flaw (-less) In diamonds flaws are imperfections, usually visible, which detract from the value of the stone; flawless diamonds are free of such defects and are highly prized
gneiss Conspicuously banded, generally coarse-grained metamorphic rock
graphite Soft black carbon stable at low pressure and temperature
igneous Describes rocks formed by crystallisation from magma
ilmenite A mineral common in kimberlite consisting of an oxide of iron and titanium and resistant to chemical alteration and decomposition during weathering
kimberlite Rock derived from deep-seated (mantle-derived) magma commonly consisting of olivine, phlogopite mica and pyroxene, often with minor ilmenite, garnet and, extremely rarely, diamond
lagerstätte An extraordinarily richly fossiliferous sedimentary deposit
lamproite Mantle-derived rock of the same general family as kimberlite but carrying minerals with slightly different compositions and some minerals not found in kimberlite, including leucite; in Australia and parts of Africa they may carry diamonds
Laurasia The Northern Hemisphere supercontinent resulting from the break-up of Pangaea, which, in turn, broke up to form the North American and Eurasian landmasses
leucite An uncommon potassium-rich mineral

found in volcanic rocks

lithosphere The surface layer of Earth, more or less rigid, comprising the crust and the upper mantle

magma Mainly molten material from which igneous rocks such as granite, dolerite and basalt crystallise

mantle The layer of rock between the crust of the Earth and its core, below a depth of approximately 35 km on the continents and 10 km in the oceans

megacryst A large crystal, commonly of garnet, thought to have been derived from near the base of the lithosphere, not related to diamonds or the kimberlite magma

Mesozoic The era containing the Triassic, Jurassic and Cretaceous Periods, from 248 to 65 million years before the present

metamorphic Describes a rock formed from a pre-existing rock of any kind which has been subjected to sufficiently changed temperature or pressure to form new minerals from the old and a new texture

mobile belt A swathe of folded metamorphic rocks usually hundreds of kilometres long and tens of kilometres wide, characterised by instability compared to the stable craton around which it is draped

olivine The most common mineral in kimberlites and some lamproites, olive green when fresh, weathering to green serpentine; the gem variety is called peridot

Pangaea A hypothetical ancient supercontinent which divided to form Laurasia and Gondwana about 300 million years ago

peridotite An ultramafic (silica-deficient) rock, dense and usually dark when fresh, the dominant rock-type in Earth's upper mantle and common as xenoliths in kimberlite

petrology The study of rocks and the conditions of their formation

phenocryst An early-formed crystal in igneous rocks, conspicuously larger than the main rock mass

picroilmenite A magnesium-rich variety of ilmenite common in kimberlite

pipe The carrot-shaped, near-surface part of a kimberlite conduit, usually becoming more linear and dyke-like at depth

plug (geological) The pipe-like upper part of a volcanic system, often harder and more erosion-resistant than the country rock into which it was intruded

pyrope garnet The common garnet of kimberlites, magnesium-rich and, in the case of G10 garnets, chrome-rich and calcium-poor; commonly ruby-red, though showing a range of colours from purple to orange

refractive index The measurement of the extent to which translucent minerals bend an entering light ray that passes through it

resorb To chemically erode a crystal that has been picked up by an ascending body of magma, the crystal not being in equilibrium with the magma, so that the resorbed crystal may change shape dramatically

rough (in diamonds) Applied to diamonds as they are recovered from the rock and until the cutting and polishing process begins

shield (geological) An area, usually large, of exposed craton

sluice (mining) An elongate box set on an incline, in which small-scale miners trap heavy minerals from a slurry of sediment and water moving over riffles in the box

spinel A family of simple oxide minerals including chromite, resistant to chemical decomposition, which may be represented in kimberlite

terrace (geological) A more or less flat, practically horizontal surface formed by rivers migrating and depositing gravel on the terrace

till Glacial outwash

tillite The rock formed when ice sheets deposit their load in a sedimentary basin

tonne A metric ton, 1000 kg

trigon An equilateral triangle that shows on the face of a rough diamond, commonly occurring in parallel clusters, the sides of the trigons parallel to the sides of the triangular crystal face of the diamond, often pointing in the opposite direction

ultramafic Igneous rocks, like peridotite, that are very silica-poor and common in the mantle

zircon The mineral, silicate of the element zirconium, which may occur in very small traces in kimberlite; it is heavier than diamond, hard and extremely stable, so barely decomposes over time

NOTES

Chapter 4

1. American Museum of Natural History website: www.amnh.org/exhibitions/diamonds/Indies.html
2. Ibid.
3. Jacques Legrand, *Diamonds: Myth, Magic and Reality*, p. 15.
4. Joan Younger Dickinson, *The Book of Diamonds: Their History and Romance from Ancient India to Modern Times*, p. 1.
5. Jacques Legrand, *Diamonds: Myth, Magic and Reality*, p. 17.
6. Ibid., p. 40.
7. A. Royden Harrison, 'The occurrence, mining and recovery of diamonds', in Diamond Drilling Symposium of the Chemical, Metallurgical and Mining Society of South Africa, April 1952.

Chapter 5

1. A.N. Wilson, *Diamonds from Birth to Eternity*, p. 242
2. Felix V. Kaminsky, Segei M. Sablukov, Elena A. Belousova, Paulo Andreazza, Mousseau Tremblay and William L. Griffin, 'Kimberlitic sources of super-deep diamonds in the Juina area, Mato Grosso State, Brazil', 9th International Kimberlite Conference Extended Abstract no. 9IKC-A-00004, 2008.
3. H.O.A. Meyer and M.E. McCallum, 'Diamonds and their sources in the Venezuelan portion of the Guyana shield', *Economic Geology*, 88, 5, August 1993, pp. 989–998.

Chapter 6

1. George Beet, *The Grand Old Days of the Diamond Fields*, p. 5.
2. Ibid., p. 50.
3. A tonne is a metric ton, a carat 0.2 of a gram, so that a grade of 0.5 cpht is equal to one tenth of a gram per tonne, or 0.0000001%, compared to an average Witwatersrand gold grade of around 10 grams per tonne, or 0.00001%, or a low-order copper mine grade of 1%.
4. Brian Roberts, *Kimberley: Turbulent City*, p. 122.
5. Ibid., p. 194.
6. Stefan Kanfer, *The Last Empire: De Beers, Diamonds and the World*, p. 179.
7. Nigel Helme, *Thomas Major Cullinan: A Biography*, p. 94.

Chapter 7

1. Olga Levinson, *Diamonds in the Desert*, p. 59.

Chapter 9

1. Chaim Even-Zohar, *From Mine to Mistress*, p. 759.
2. M.B. McClenaghan and B.A. Kjarsgaard, 'Indicator mineral and surficial geochemical exploration methods for kimberlite in glaciated terrain, examples from Canada', in *Mineral Resources of Canada: A Synthesis of Major Deposit-types, District Metallogeny, the Evolution of Geological Provinces and Exploration Methods*, Geological Association of Canada, Special Paper 5, pp. 983–1006.
3. Kevin Krajick, *Barren Lands: An Epic Search for Diamonds in the Canadian Arctic*, pp. 163–373.

Chapter 10

1. The remarks are from Professor Rex Prider's introductory address at the Fourth International Kimberlite Conference in Perth in 1985.

Chapter 12

1. R.J. Rayner, M.K. Bamford, D.J. Brothers, A.S. Dippenaar-Schoeman, I.J. McKay, R.G. Oberprieler and S.B. Waters, 'Cretaceous fossils from the Orapa diamond mine', *Palaeont. Afr.*, 33, 1997, pp. 55–65.
2. Stephen Jay Gould, *Wonderful Life: The Burgess Shale and the Nature of History*, pp. 23–32.
3. www://geokhi.ru/~meteorit/opis/novo-urei-e.html
4. www://meteorite.fr/en/classification/PAC-group.htm
5. Jozsef Garai, Stephen E Haggerty, Sandeep Rakhi and Mark Chance, 'Infrared absorption investigations confirm the extraterrestrial origin of carbonado diamonds', *Astrophysical Journal*, 2006, p. L153.

Chapter 13

1 M.C.J. de Wit, 'Cainozoic evolution of drainage systems in the north-western Cape', PhD thesis, University of Cape Town, 1993.
2 B.J. Bluck, J.D. Ward and M.C.J. de Wit, 'Diamond mega-placers: southern Africa and the Kaapvaal craton in a global context', *Mineral Deposits and Earth Evolution*, Geological Society, London, Special Publications 248, 2005, p. 213.
3 R.I. Spaggiari, B.J. Bluck and J.D. Ward, 'Characteristics of diamondiferous Plio-Pleistocene littoral deposits within the palaeo Orange River mouth, Namibia', *Ore Geology Reviews*, 28, p. 476.
4 I. Corbett, 'The sedimentology of the diamondiferous deflation deposits, Namibia', PhD thesis, University of Cape Town, 1989.
5 D. Phillips and J.W. Harris, 'Diamond provenance studies from $^{40}Ar/^{39}Ar$ dating of clinopyroxene inclusions: an example from the west coast of Namibia', *Lithos*, 2009, doi:10.1016/j.lithos.2009.05.003, pp. 1–12.

Chapter 14

1 Colin Newbury, 'The origins and function of the London Diamond Syndicate', *Business History* 29 (1), pp 5-26, 1987

SELECT BIBLIOGRAPHY

Angove, John. *In the Early Days: The Reminiscences of Pioneer Life on the South African Diamond Fields*. Kimberley: Handel House, 1910

Balfour, Ian. *Famous Diamonds*, 2nd edn. Santa Monica: Gemological Institute of America, 1992

Beet, George. *The Grand Old days of the Diamond Fields*. Cape Town: Maskew Miller, c.1931

Blakey, George G. *The Diamond*. London: Paddington Press, 1977

Bruton, Eric. *Diamonds*, 2nd edn. London: N.A.G. Press, 1978

Burgess, P.H.E. *Diamonds Unlimited*. London: John Long, 1960

Chilvers, Hedley A. *The Story of De Beers*. London: Cassell, 1939

Du Plessis, Captain J.H. *Diamonds Are Dangerous*. New York: John Day, 1961

Edwards, Hugh. *Kimberley: Dreaming to Diamonds*. Perth: Tangee Publishing, 2005

Epstein, Edward Jay. *The Rise and Fall of Diamonds*. New York: Simon and Schuster, 1982

Even-Zohar, Chaim. *From Mine to Mistress*. London: Mining Communications, 2007

Fleming, Ian. *The Diamond Smugglers*. London: Pan Books, 1960

Garavini di Turno, Sadio. *Diamond River*. London: Hamish Hamilton, 1963

Green, Lawrence. *So Few Are Free*. Cape Town: Howard B. Timmins, 1946

Hahn, Emily. *Diamond*. London: Weidenfeld and Nicolson, 1956

Harbottle, Michael. *The Knaves of Diamonds*. London: Seeley Service, 1976

Hart, Matthew. *Diamond: The History of a Cold-blooded Love Affair*. London: Fourth Estate, 2002

Helme, Nigel. *Thomas Major Cullinan: A Biography*. Johannesburg: McGraw-Hill, 1974

Herbert, Ivor. *The Diamond Diggers: South Africa 1866 to the 1970s*. London: Tom Stacey, 1972

Idriess, Ion L. *Stone of Destiny*. Sydney: Angus and Robertson, 1948

Jessup, Edward. *Ernest Oppenheimer: A Study in Power*. London: Rex Collings, 1979

Kamil, Fred. *The Diamond Underworld*. London: Allen Lane, 1979

Kanfer, Stefan. *The Last Empire: De Beers, Diamonds and the World*. London: Hodder and Stoughton, 1993

Koskoff, David E. *The Diamond World: The True Behind-the-Scenes Story of the Mysterious Circle*

That Trades in the Most Spectacular Gem. New York: Harper and Row, 1981

Krajick, Kevin. *Barren Lands: An Epic Search for Diamonds in the North American Arctic*. New York: Henry Holt, 2001

Kurin, Richard. *Hope Diamond: The Legendary History of Cursed Gem*. New York: Harper Collins, 2006

Legrand, Jacques. *Diamonds: Myth, Magic and Reality*. New York: Crown, 1980

Lehmann, Olga. *Look beyond the Wind: The Life of Dr. Hans Merensky*. Cape Town: Howard Timmins, 1955

Levinson, Olga. *Diamonds in the Desert*. Cape Town: Tafelberg, 1983

Loring, John. *Tiffany Diamonds*. New York: Harry N. Abrams, 2005

Machens, Eberhard W. *Platinum, Gold and Diamonds: The Adventure of Hans Merensky's Discoveries*. Pretoria: Protea Book House, 2009

MacNevin, A.A. *Diamonds in New South Wales*. Department of Mines, Geological Survey of New South Wales, Sydney, 1977

McCarthy, James Remington. *Fire in the Earth: The Story of the Diamond*. London: Robert Hale, 1946

McCarthy, Terence and Bruce Rubidge. *The Story of Earth and Life: A Southern African Perspective on a 4.6-billion-year Journey*. Cape Town: Struik, 2005

Meredith, Martin. *Diamonds, Gold and War: The Making of South Africa*. London: Simon and Schuster, 2007

Monnickendam, A. *The Magic of Diamonds*. London: Hammond, Hammond, 1955

Mouawad-Bari, Hubert and Violaine Sautter. *Diamonds: In the Heart of the Earth, in the Heart of Stars, in the Heart of Power*. Paris: Vilo and Adam Biro, 2001

Roberts, Brian. *The Diamond Magnates*. London: Hamish Hamilton, 1972

Robertson, Marian. *Diamond Fever*. Cape Town: Oxford University Press, 1974

Simons, Phillida Brooke. *Cullinan: Diamonds, Dreams and Discoveries*. Cape Town: Fernwood Press, 2001

Wagner, P.A. *The Diamond Fields of Southern Africa*. Cape Town: C. Struik, 1971

Wharton-Tigar, Edward, and A.J. Wilson. *Burning Bright: The Autobiography of Edward Wharton-Tigar*. London: Metal Bulletin, 1987

Williams, Alpheus F. *The Genesis of the Diamond*. London: Ernest Benn, 1932

Williams, Gardner F. *The Diamond Mines of South Africa*. New York: B.F. Buck, 1905

Williams, Roger. *King of Sea Diamonds: The Saga of Sam Collins*. Cape Town: W.J. Flesch, 1996

Wilson, A.N. *International Diamond Annual*, vol. 1. Johannesburg: Diamond Annual, 1971

Wilson, A.N. *A Review of the World's Diamond Industry and Trade*. Johannesburg: Diamond Annual, 1972

Wilson, A.N. *Diamonds: From Birth to Eternity*. Santa Monica, California: Gemological Institute of America, 1982

Worger, William H. *South Africa's City of Diamonds: Mine Workers and Monopoly Capitalism in Kimberley, 1867–1895*. New Haven: Yale University Press, 1987

INDEX

PHOTO CREDITS

De Beers Consolidated Mines
7, 10, 13, 14, 16, 22, 33, 35, 36, 37, 39, 43, 45, 93, 99, 101 (bottom), 103 (top and bottom), 105, 107, 109, 113, 114, 115, 122, 123, 125, 128, 129, 134, 136, 138, 139, 147, 148, 160, 164, 173, 174, 176, 179, 184, 208, 209, 221, 224, 225, 226, 229, 248, 258, 268, 270, 273, 275, 280, 286, 287

Gem Diamonds
21 (bottom), 28, 29, 34, 119, 141, 142, 144, 146, 207, 210, 222, 252, 274

Rio Tinto/Rio Tinto Diamonds
19, 65, 66, 182, 183, 186, 190, 202, 204, 205, 276, 278

Brenthurst Library/ Clive Hassall
17, 96, 98, 106, 271, 288

Thomas Branch/Mineral Services
21, 230, 231, 232, 236, 238, 279

Trish Heywood
127, 272

Trish Heywood/Chris Hellinger
131

Trish Heywood/UCT Geology Department
228

Trish Heywood/Mineral Services
237

Luiz Claudio Marigo
70, 75, 198

Alrosa
159, 155, 157 (top), 162, 163

Nick Norman
18, 26, 31, 41, 83, 97, 101, 214, 219, 220, 251, 266

Nick Norman/Maggie Newman
259

Hans Merensky Holdings
112, 261

Hans Merensky Holdings/ Clive Hassall
126, 253

Sunday Times
282

Ian Fleming, *The Diamond Smugglers*
284

Ian Corbett
264

Mike de Wit
218, 256

Petra Diamonds
41 (bottom), 250 (bottom)

Keith Whitelock
143

Victor Masaitis
158, 243

Bernard Price Institute for Palaeontology
241

John Gurney/Mineral Services
233

Chris Smith
199

Courtesy of Rebecca Hudson, Melbourne, Australia
191

Cluff Resources Pacific NL
189

Edward Gajdel Photography Inc. 2010
180 (bottom)

Chris Jennings
177, 180 (top)

Geological Association of Canada
172

Bill Larson, Pala International, California
167

Vladimir Shchukin
157 (bottom)

De Beers: 1888-1988, Société Général de Belgique and Sibeka
120

John Lincoln
48, 110

Albany History Museum, Grahamstown H37/1
90

Vaaldiam Mining Inc., Toronto
2, 87

Harrison Cookenboo
86

Guilherme Gonzaga
85

Leila Benitez
80

A. Sthapak
64

Nick Sobolev
44

Jay Barton
27

Dr Tom Clifford
25

Missouri Botanical Garden
23

Gallo Images/Getty Images
78, 170, 198 (bottom), 201, 240, 247, 269

Gallo Images
42, 260

The Bigger Picture/Alamy
40, 47, 77, 89, 130, 192, 195

The Bigger Picture/ Reuters
150, 169, 285

Topfoto/INPRA
12, 59, 60, 63, 73, 74

Rex Features/INPRA
281

Greatstock/CORBIS
46, 69, 108, 244, 245

Bridgeman Art Library
57, 61

akg-images
54

British Library X6576
50

VincentGrafhorst/ Afripics.com
216

Brenda Veldman
255

Johann van Tonder/ PictureNET Africa
215

Ruvan Boshoff/Sunday Times
263

John Sparks/Nature PI/Digital Source
154

Robert Valentic/Nature PI/ Digital Source
196

Luiz Claudio Marigo/Nature PI/Digital Source
84

Christoph Becker/Nature PI/ Digital Source
265

First published by Jacana Media (Pty) Ltd in 2010

10 Orange Street
Sunnyside
Auckland Park 2092
South Africa
(+27 11) 628-3200
www.jacana.co.za

ISBN 978-1-77009-828-2

Cover and text design and layout by Abdul Amien
Set in Minion
Printed by Craftprint, Singapore
Job no. 001297

See a complete list of Jacana titles at www.jacana.co.za